Organizational Excellence

This book discusses the multifaceted nature of organizational excellence and proceeds to consider the roles played by data, technological advances leading to innovation and value-based leadership in guiding an organization in its journey towards excellence. Moving beyond conventional growth metrics, this book presents excellence as a continuous journey of improvement, integrating data-driven decision-making, digital transformation, and value-based leadership. The book highlights the transformative role of AI and advanced digital technologies in optimizing processes, fostering innovation, and enhancing performance measurement. It also explores Six Sigma, Lean Management, reviews frameworks like Baldridge, EFQM, Excellence Canada, and other quality improvement frameworks, emphasizing the need for continuous process refinement. It is an essential resource for leaders, managers, and professionals in management, strategy, operations, and innovation, as well as scholars in management science and organizational development.

This book accomplishes the following:

- Highlights the role of data analytics and AI in process optimization and innovation; explores the critical factors driving sustained organizational success
- Introduces scenario planning and robust decision-making to develop strategies
- Distinguishes excellence from growth and presents it as a continuous improvement journey
- Considers institutionalized innovation and new product development as essential
- Covers Six Sigma, Lean Management, and other methodologies, emphasizes governance, ethics, and corporate social responsibility.

Artificial Intelligence, Machine Learning, Data Analytics and Automation for Business Management

Rajesh Dey, Senior Commissioning Editor, CRC Press, Taylor & Francis Group

The series presents a comprehensive exploration of how organisations strategically leverage technology to achieve their goals and maintain competitiveness in rapidly evolving industries. With contributions from experts across multiple disciplines, including computer science, business management, and data science, the series adopts an interdisciplinary approach, offering diverse perspectives and fostering cross-functional collaboration. In an era defined by rapid technological innovation and digital transformation, the book series on Computational Intelligence, Data Analytics, and Emerging Technology for Business Management serves as an indispensable resource for professionals, researchers, and students seeking to harness the power of technology to drive organisational growth and success.

- Includes works on the latest advancements in computational intelligence, data analytics, and emerging technologies to offer invaluable insights into leveraging cutting-edge tools and methodologies to drive organisational success.
- Emphasizes practical applications, featuring real-world case studies, examples, and best practices that demonstrate how these technologies can be effectively implemented to address business challenges and drive innovation.
- A multidisciplinary series draws on principles from management, engineering, economics, and other disciplines to effectively leverage technology for organisational success.
- Comprising professional insights, reference materials, and case studies, this series delves into key concepts essential for effective technology for management success.
- Serves as an indispensable resource for scholars, practitioners, and decision-makers committed to driving innovation and sustainable growth in the digital age.

Green Engineering for Optimizing Firm Performance: AI and Automation for Sustainable Technologies
By Sonal Trivedi, Balamurugan Balusamy, Liza Macasukit Gernal, Mahmoud Ahmad Al-Khasawneh

Organizational Excellence: Data, Technology and Leadership
Shyama Prasad Mukherjee

www.routledge.com/ *Artificial-Intelligence-Machine-Learning-Data-Analytics-and-Automation-for- Business-Management/AIMLDAABM*

Organizational Excellence

Data, Technology and Leadership

Shyama Prasad Mukherjee

CRC Press
Taylor & Francis Group
Boca Raton London New York

CRC Press is an imprint of the
Taylor & Francis Group, an **informa** business
A CHAPMAN & HALL BOOK

First edition published 2025
by CRC Press
2385 NW Executive Center Drive, Suite 320, Boca Raton FL 33431

and by CRC Press
4 Park Square, Milton Park, Abingdon, Oxon, OX14 4RN

CRC Press is an imprint of Taylor & Francis Group, LLC

Library of Congress Cataloging-in-Publication Data
Names: Mukherjee, S. P. author
Title: Organizational excellence : data, technology and leadership /
Shyama Prasad Mukherjee.
Description: First edition. | Boca Raton, FL : CRC Press, 2025. |
Series: Artificial intelligence, machine learning, data analytics and automation
for business management | Includes bibliographical references and index.
Identifiers: LCCN 2025015663 (print) | LCCN 2025015664 (ebook) |
ISBN 9781032970202 hardback | ISBN 9781032970226 paperback |
ISBN 9781003591733 ebook
Subjects: LCSH: Management–Statistical methods |
Technological innovations–Management | Organizational effectiveness
Classification: LCC HD30.215 .M845 2025 (print) | LCC HD30.215 (ebook) |
DDC 658.4/033–dc23/eng/20250612
LC record available at https://lccn.loc.gov/2025015663
LC ebook record available at https://lccn.loc.gov/2025015664

ISBN: 9781032970202 (hbk)
ISBN: 9781032970226 (pbk)
ISBN: 9781003591733 (ebk)

DOI: 10.1201/9781003591733

Typeset in Times
by Newgen Publishing UK

Contents

Preface

More than two decades back I received an invitation from the Asiatic Society, Calcutta to deliver a Memorial Lecture primarily meant for its members and associates. By that time, I came to be familiar with the performance of a wide variety of organizations ranging from private manufacturing industries through institutions of higher education to publicly funded Research and Development establishments and scientific societies and associations. And I decided to speak on Organizational Excellence. My involvement in a wide spectrum of organizations, mostly in an advisory capacity and rarely in an executive one-had led me to think about enablers and outcomes in the context of organizational performance and to go through the extant literature on the subject. I have been actively involved in the movement for Quality and Productivity through my academic pursuits and I found a new avenue to continue with those. The present book is a reflection of that. No doubt, my academic and professional background influenced the presentation of my thoughts and ideas on a subject that has attracted the attention of many reputed academicians and professionals or even practitioners.

The present volume is meant to provide a concise narrative on organizational excellence, focusing on both process performance as also on the outcomes achieved, in an ongoing exercise towards better performance. Starting with an attempt to distinguish 'excellence' from correlates like success, growth, competitive advantage, perfection and the like, the book builds on the notion that excellence is a journey rather than an end. This journey is characterized by a carefully crafted vision and mission, a robust strategy based on sound scenario analysis, a conducive work culture, business processes adapted to contemporary advances in technology, a strong focus on innovation, a resilient workforce, and a value-based leadership.

The content includes a comprehensive discussion on measuring organizational performance not merely for traditional business organizations but also for academic institutions and R & D establishments, Business Excellence Models have been briefly but critically reviewed and some open problems in using any such model have been hinted at. In its treatment of the journey to excellence, the author has considered the roles of change management along with the extant improvement methodologies. An attempt has been made to indicate the role of Innovation in this context and some broad guidelines for institutionalization of innovation have been provided. The over-arching role of continuous improvement using contemporaneous developments in mapping, analysis and optimization of business processes has been highlighted. A balanced strategy has been advocated for ethical and responsible use of technology, particularly generative AI, in which technological advancements are infused with fundamental ethics and values.

This volume is not intended to be a text -book that can be readily prescribed to students attending a formal course on management or on organizational development. It can modestly claim the status of a reference book for serious students and research workers as well as professionals interested in the subject, which is of great contemporary interest. Readers will find some materials bearing upon topics like decision-making, scenario analysis, business process re-engineering, change management, theory of inventive problem-solving, technological innovation, generative artificial intelligence and the like that do not usually find a place in the context of a discussion on organizational excellence.

Those interested in the subject will find some fresh ideas and thoughts in the pages of this book, which may initiate new vistas for discourses on organizational excellence, not restricted to traditional business organizations. Clarity of thoughts and ideas and simplicity and brevity in presentation of the content will be hopefully appreciated by the readers.

In preparing the manuscript, the author received constant encouragement and help from Rajesh Dey, the Commissioning Editor, which deserves due acknowledgement. The manuscript could be completed only through the technical help and support from Munna Saha, a young professional, for preparing the tables and the diagrams besides incorporating changes in the manuscript on lines suggested by the Copyeditor. Thanks are also due to Kari Budyk, the Production Editor, and Surya Rajasekar, the Project manager, for their help and guidance. I will be failing in my duty if I do not express my gratitude to my wife Reba for the continued support she extended right from the beginning.

Shyama Prasad Mukherjee
April 2025

Author Biography

Shyama Prasad Mukherjee (1938–) started teaching statistics in the Presidency College, Calcutta, in 1962. He joined the University of Calcutta in 1964 as a Lecturer and became Centenary Professor of Statistics in 1982. After retirement in 2004, he remained a Visiting Professor till 2012. He has published around 120 research papers in reputed journals like *Journal of Multivariate Analysis*, *Journal of Applied Probability, Naval Research Logistics Quarterly, Communications in Statistics, Sankhya, IEEE Transactions on Reliability, Calcutta Statistical Association Bulletin, Journal of the Indian Society of Agricultural Statistics, Journal of Indian Statistical Association, Opsearch, Micro-electronics & Reliability, International Journal of Probability & Statistics*, among others. He has also published a host of expository and review articles and has authored several books and edited a few others. Dr. Mukherjee has guided 25 research students in statistics, mathematics, statistical and probabilistic aspects of management and engineering. He is the Chairman of the Board of Directors of the International Statistical Education Centre (ISI) and Mentor in the Indian Association for Productivity, Quality & Reliability. He was the President of the Indian Society of Probability and Statistics and Operational Research Society of India, and Vice-President of the International Federation of Operational Research Societies (IFORS). He has delivered lectures and conducted short courses in many universities in Asia, Europe and North America.

1 Introduction

1.1 PROLOGUE

The extant volume of published materials related to organizational excellence, such as books and monographs, review and research articles as well as reports, is quite rich. Communality in thoughts and ideas reflected in statements, elucidations and illustrations is just expected. However, from purely abstract philosophical underpinnings, on the one hand, to concrete real-life realization on the other, organizational excellence presents a big canvas. This manifests in the diversity of definitions, approaches and methods adopted by different contributors. Given this context, any addition to the existing literature possibly owes an explanation. Concepts like honesty or beauty, credibility or resilience, commitment and devotion, ethics or values and the like defy unanimous views, consensus or even widely accepted definitions.

These are innate virtues or attributes, which may be cultivated and elevated through internal efforts. However, these are perceptible though not necessarily obviously visible from outside. Qualifying concepts like excellent or superb, brilliant or exceptional, significant or remarkable and similar others that indicate the level of achievement and/or manifestation of some virtue or attribute are also difficult to define. Keeping aside the objectivity–subjectivity debate, it can be asserted that some amount of subjectivity leads to the absence of unique definitions of such concepts.

While the absence of widely accepted generic definitions is recognized, attempts have been made to 'assess or judge' the degree or level of any such attribute inherent in or/and exhibited by an individual or an organization to suit specific situations and requirements. In fact, there has been a growing tendency to 'quantify' these concepts and 'measure' the associated degrees or levels. Such measures enable the use of relevant quantitative tools for analysis and inference.

It is easier to arrive at an agreed definition of an 'organization' than to attempt a definition of 'excellence'. Of course, the term 'organization' spans a big canvas, as it can be applied to agricultural farms and manufacturing firms, through public and private providers of various services, to purely academic institutions and research and development organizations. This diversity in the nature of organizations is obviously associated with a corresponding diversity in their objectives and goals reflected in their core and support processes, and organizational excellence holds relevance in such processes.

DOI: 10.1201/9781003591733-1

Excellence could relate to process performance or output performance, or both taken together. Experiences reveal the existence of situations where an excellent performance of different processes carried out internally by an organization is not positively or directly reflected in the results, which are quite likely to be influenced by externalities. Unless duly adjusted for such external influences—a task quite difficult to achieve—the results may not truly characterize the organization's performance. An obvious consequence has been a polarity in the concept of 'excellence', one focusing on standards, processes and procedures within the organization, while a comprehensive and consolidated account of results achieved is used to assess performance. In fact, the present author relies on the model of 'quality of performance' as the composite of 'quality of design' and 'quality of conformance (to design requirements)' for an understanding of performance-related excellence. Excellence may be interpreted in absolute or relative terms. Used in a relative sense, excellence is judged in terms of quality of conformance and even in terms of quality of performance, given the design meant to achieve certain specified goals and objectives. The advocates of the absolute interpretation would argue that excellence means striving to attain better-than-designed targets.

Accepting the idea that excellence is an attribute of performance and recognizing the fact that the performance of any organization cannot be assessed in terms of a single metric, we need a set of metrics to characterize organizational performance. The choice of the metrics depends on the type of the organization and can hardly be unique. While the performance of each individual worker in the organization contributes to the overall performance of the organization, however, the latter cannot be the sum of these individual contributions judged by any metric. And there exist several aspects of the environment that influence the overall performance and possibly not the performance of any one individual. Developing an appropriate multiple measure of organizational performance is by itself a challenging task.

There has been a growing tendency to use generic approaches, measures and tools that are claimed to enjoy universal applicability, without focusing on specific features of individual entities or types of similar entities. This tendency, which is a fallout of the globalization phenomenon, has its merits and limitations. It highlights even conceited or artificial similarities at the cost of genuine and important specificities. The measures of organizational performance and yardsticks to compare them against the goals and objectives in a charity trust cannot be the same as those for an educational institution, a business house or a hospitality industry. The measures used by an organization to assess its performance may not always consider the factors of the 'environment', which focus on 'what is right for the society' and not simply 'what is good for the organization', while interacting with or overseeing the organization. This highlights the role of leadership and values and ethics in achieving excellence. In fact, some contributors to the field discuss the role of value-based leadership in the context of organizational excellence.

An important question like 'Does growth necessarily imply excellence?' or its reverse 'Does excellence necessarily imply growth?' usually gets an equivocal answer. Similarly, it is difficult to get a consensus on considering excellence as a level of performance during a certain period or in terms of a continuous and consistent effort to excel over peers and competitors—at least some of them—over a reasonably long period. In other words, we should explore the nature of relationship between excellence and sustainability of performance. While a general tendency to associate 'competitive edge or advantage' with organizational excellence is articulated by some thought leaders and professionals, there are others who argue that there exists no necessary relationship between the two. Excellence can be spoken of in relation to a monopolistic organization and with peers and competitors enjoying some unique locational or other external advantage; an organization seriously pursuing the path of excellence may not necessarily exhibit a competitive advantage.

The virtual deluge of definitions and connotations of 'organizational excellence' is quite expected. Equally problematic is developing a holistic characterization of organizational excellence that embodies a comprehensive description of organizational performance, followed by an attempt to measure it and developing a model in terms of its factors or drivers, to control or stabilize it at a desired level, and finally a road to consistently improve upon the existing level. It's little surprising that a whole host of 'critical success factors' related to organizational excellence and correspondingly a growing set of models to explain organizational excellence have surfaced over the years.

Remarkable advances in technology, including those in digital technology, have taken place over the last few decades to shake the old world order in every field of human activity. Therefore, a discussion on organizational performance and organizational excellence today cannot be meaningful without thoughts and ideas about harnessing these advances. While details of such development need to be fully comprehended by a selected few, its applications and consequences should engage the attention of all.

The present chapter contributes to an understanding of these concepts, steering clear of controversies and aiming at simplicity, rigour and self-completeness. Some of the ideas and thoughts presented herein may simply reflect the extant views, while a few others may hopefully enrich the current ideas. The presentation is intentionally in a narrative form, and not much technical. The motivating principle has been mainly simplicity, consistency and amenability to analytical treatment.

1.2 ORGANIZATIONS

Organizations are better understood than defined. We can settle for some definition that arises from this understanding.

An organization is an entity (often a social unit) comprising multiple persons, such as an institution or an association, that has or works toward a collective goal and is linked with an external environment. The environment affects it and is affected by it.

Organizations include individuals engaged at different levels of responsibility and authority in different activities carried out in different ways to achieve some common goals and objectives. Some extraordinary individuals are considered as institutions by themselves. However, such individuals are not by themselves organizations; they lead and guide organizations where others also are involved. Organizations are often known for their leaders.

Datta (2019) elaborated on the ideas of the great Indian saint Swami Vivekananda about 'organization' in a somewhat unconventional manner. He first expanded each of the 12 letters (some repeated twice) in 'organization' in the following manner:

O stands for Obedience among members to leaders of the organization
R stands for Re-enforcement to unite members into a single entity at least for a specified purpose symbolized by the organization
G stands for Goal which the organization should strive to achieve, concretized in terms of objectives and targets
A stands for Acquisition and Accumulation of knowledge and skills by members of the organization collectively to form the knowledge and skill base of the organization
N denotes Network among members, spelling out interactions and lines of command
I implies an Ideal for harmony and mutual respect among members
S indicates Structure, giving out responsibility and commensurate authority for each member, avoiding overlaps and conflicts
A stands for Attention to be paid constantly to the goals and priorities of the organization so that deviations and distortions can be detected and removed as quickly as possible
T denotes Thoughts, implying that members must be constantly thinking about improving their pursuit of the ideal and the goal of the organization
I stands for Impartiality in thoughts and deeds of all members so that the organization is not misunderstood to carry any bias towards any particular view of the external world
O stands for Oblation or submission of members to the organization's ideal as a religious offering and
N signifies Nation-building as the eventual goal of the organization

While the above elucidation may appear to be a bit philosophical and idealistic, the features of an organization as spelt out here are all relevant—maybe to different extents in different types of organizations—and none of these should be regarded as redundant or useless. No doubt, the above elucidation applies more to non-commercial or charitable service organizations or groups of people engaged in humanitarian work.

A wide range of organization types exists from charitable trusts and not-for-profit associations, to educational institutions and research laboratories, government agencies, manufacturing and trading houses, religious missions to even political parties. Even within the type branded as educational institutions, relevant differences exist among different levels and forms of education and between private and public organizations. Similar is the case with research and development organizations. Some organizations may be registered with some statutory bodies under some regulatory framework, while others are not. Proprietary or partnership organizations, companies with limited liability, self-help groups, statutory bodies and public sector service enterprises thus give rise to the wide diversity of organizations.

Well-defined structures exist in some organisations, while some others just loosely bind their constituents together. Constituents of organizations may have their respective avocations, personal commitments, interests, attitudes and aptitudes. However, they are all united to work under the same leadership, within the same organizational value system and for the same purpose (as spelt out in the general goals and objectives of the organization and the roles and responsibilities of specific teams and individuals).

'Environment' in this context combines all the concerns of all the stakeholders in the activities of an organization and the impacts thereof on the physical as well as the socio-economic environment. It includes consumer bodies, regulatory agencies, civil societies and all other bodies which monitor the organization's activities and their policies and plans. Environment obviously includes changes in weather conditions, which can affect employees' engagement in organizational activities. Environment does imply both constraints and impediments as well as advantages and a push for improved performance.

Organizations can be gainfully likened to systems. A system in system science is a set of inter-related elements (branded in the present context as branches departments or functional areas or groups or teams) possessing the following four characteristics:

1. *Content*: implying ability of the system to identify its goals and objectives
2. *Structure*: implying differentiation among the elements in terms of their processes and responsibilities, as also among units at different levels within each element in terms of their responsibilities and authorities
3. *Communication*: implying the nature, mode and frequency of interactions among different elements, between an element and the external environment, and between the system as a whole and the external environment and
4. *Control*: implying the ability of the system to modify or change some of the above three characteristics as and when found necessary (through inspection, audit, review and the like).

Elements within a system may be strongly or loosely interrelated to one another. It is quite possible to consider the aggregate of all organizations working under the same broad umbrella of a well-defined set of goals and objectives, a common overarching management and similar structures as a system. Thus, we speak of a healthcare delivery system, a school education system or a public distribution (of essential food items) system. Maybe, we can recognize the health care or school education services provided by units in the public and the private sectors as two different systems. In such a framework, each of the organizations within or under a system can be regarded as one element of the system. The only point to take note is that elements in such a system are relatively loosely knit, compared to those within a system corresponding to a single organization.

In fact, recent studies on organizational development emphasize that an organization (irrespective of its domain of operation or its ownership) should be rather treated as an organic system possessing capabilities—especially to adapt and adjust—that mark living organisms. These capabilities are of crucial importance for any organization to navigate its business processes through challenging and even threatening externalities that are becoming the order of the day. These capabilities should be duly harnessed to guide an organization on its intended path to excellence.

While there is no need to distinguish not-for-profit organizations from organizations meant to earn profits or organizations in the private sector from those in the public sector in a discussion on organizational excellence, a distinction between a stand-alone organization competent and fully authorized to set its priorities and preferences (and that way to formulate it policy and strategy) and an organization under the umbrella of an overarching organization providing the broad framework for its operational policies and practices (and regulates the provisioning of inputs) remains important. Of course, the concept of excellence or the march towards excellence is not all irrelevant to an organization in the second category. In fact, one comes across public service organizations operating under the broad guidance of a government department and providing noticeably satisfactory services to its clients or beneficiaries. We remain confined to stand-alone organizations and organizations fully authorized to chalk out their policies and strategies as well as change them, as and when deemed necessary by the organization, by its top leadership.

Before closing this section on organizations, it may be worthwhile to consider an organization as a team of individuals or groups with many hands, but one head. This head does not symbolize top leadership with the titles of CEO, CMD or MD, but signifies the organ that leads members to think in unison. A successful team may be taken as representing the situation where 'together (T) everyone (E) achieves (A) more (M)'. No element that deters the organization from achieving what it deserves in the environment in which it operates characterizes the rationale of organizations for the purpose of our discussions.

1.3 ORGANIZATIONAL PERFORMANCE

'Performance' may relate to (1) activities carried out or processes performed by an organization or (2) outcomes or outputs or impacts of such activities or processes in relation to the overall goals and objectives to be achieved by the organization. A critical approach to the operationalization of the concept has to carefully take into account measures of individual performance to be incorporated in the superstructure for assessing organizational performance. Further, a superstructure by itself may be a philosophical guide and may not work as a concrete method to link the two. There exist views that organizational performance depends more on externalities rather than internal processes and individual contributions to such processes.

Processes include both core and support activities and encompass people management, resource management and interactions with the environment. Outcomes could include financial results, stakeholder satisfaction, impact on society and impact on the environment. Each of these outcomes can be judged in terms of multiple measures and may require disaggregation to desired degrees. Thus in the assessment of stakeholder satisfaction, separate measures may have to be developed and used for employees, investors, regulatory bodies and people's representatives. The multi-dimensional nature of organizational performance has been duly brought out in the much-discussed balanced score card (which is not just a single score), which is known to balance (and not just to combine) the leading variables associated with various processes with lagging variables reflected in different results that follow, such as financial and non-financial results, present and future scenarios, and internal and external views about the organization's performance.

Not too often, different performance parameters of any organization reveal some form of inconsistency or conflict in the sense that one parameter may behave in one direction and a second in the reverse direction with a particular policy, strategy or operational tactics. Trade-offs are sometimes adopted and weights assigned to different parameters vary. While the assignment of weights to measures of performance in different dimensions is not an easy task, some sensitivity studies should be carried out before taking a weighted aggregate or average of the measures on different fronts.

With emphasis placed on sustainability of performance in order to rate the organizational performance as excellent or otherwise, it becomes imperative to choose such measures for any dimension of performance to reflect the trend in performance and the capacity to retain the same over a reasonably long period to time. Scarcity of data, especially in the case of organizations that have started more recently to function, at least under the current leadership, may stand in the way. The predictive value of any measure should also be duly noted.

In the case of academic institutions (particularly those involved in imparting education at the tertiary level) as well as research and development organizations, we note that the processes performed and results or outcomes achieved by such institutions differ markedly from those applicable to usual business organizations. In fact, the profile of stakeholders, including direct and indirect beneficiaries, as well as the needs and expectations of such stakeholders may vary significantly between a manufacturing or trading organization and a service organization in the public sector. We note that for such organizations, performance is often judged relative to certain frameworks that are essentially meant to rank such organizations according to some interpretation of 'excellence'. And in several such ranking frameworks, opinions of experts and stakeholders form a part of the assessment process. In the case of R & D organizations, the task is quite complicated and context-specific measures of performance should be considered.

1.4 EXCELLENCE

Excellence does not lend itself to a solid definition, not to speak of a consensus measurement. Notwithstanding this problem, most individuals and organizations would like, and even, strive to achieve excellence. Excellence goes beyond effectiveness and efficiency but may not necessarily imply growth—howsoever we characterize the latter.

Among many definitions of excellence, one that is somewhat concrete could be stated as: excellence is an outstanding practice (that stands out distinctly from others) in managing the organization (essentially its operations) and in achieving the desired results (as spelt out in its goals and objectives and as integrated in terms of some trade-offs to take care of complementarity and conflict among different objectives)—a practice that need not be absolutely the best but is worthy of emulation by others. Thus, excellence applies to the execution of different operations and their results. Of course, the results depend on the effectiveness and efficiency in execution.

Excellence implies staying ahead of the pack, outperforming competitors in the considered area. Excellence implies standardization, when it is needed, but it mainly implies positive differentiation. In fact, standardization applies to everyone, while differentiation is restricted to a few who are able to make it. Differentiation in organizational performance can lead to excellence; but excellence is where the ladder ends, but it not the ladder itself. The focus should be placed on the steps in the ladder and how well do we climb these steps. These steps relate to the organization's internal processes and tasks performed by its people by acquiring and deploying requisite resources—physical, material, technological as well as human (knowledge and skills of the workforce).

Excellence has been sometimes rightly interpreted as a continuous effort to reach an ambitious goal, rather than the goal itself. Since 'excellence' implies excelling beyond simply meeting stakeholder needs (as expressed) and expectations (as remaining at the back of their minds). Excelling also implies outdoing or surpassing peer organizations (to avoid the unwarranted term 'competitors'). Since the other organizations—at least some of those—will be striving to improve their current levels of performance, excellence demands the organization to continuously achieve higher levels of performance.

The most important stakeholder group which justifies the existence of any organization is the group generically addressed as 'customers'. In fact, this term has different connotations in different types of organizations. Even for a manufacturing organization, the group of customers having direct interface with the supplier or provider organization may have requirements related to delivery, shelf-life and storage, or return of unused or defective supplies, for example. The more important group of end-users who usually do not have a direct interface with the producer may have requirements with regard to product quality and reliability, maintenance and disposal, for example. For a private service organization, users do generally have a direct interface with the provider. For public sector service organizations, some services may have to be provided to all in a specified group proactively without the beneficiary or potential beneficiary demanding it. The municipality is an example of this kind of public service organization. For the education sector, direct beneficiaries are the students, but stakeholders are the guardians and parents of these students, employers including the government and even the society at large. Their expectations are not directly communicated to the organizations imparting and managing education. In the case of R&D organizations, clients do exist if the research or development work has been sponsored by some business house. In the case of publicly funded organizations, beneficiaries and potential beneficiaries have to be identified with due care and diligence for each case separately.

Excellence has been often linked—may be implicitly, if not in a covert manner—with 'competition'. It may not be wise to completely ignore this linkage. However, the initiative to follow the road to excellence should not originate from a spirit of competition, which, unfortunately in some cases, tends to deviate from professional ethics and recognized value systems. It may not be out of place to mention that organizations can and do achieve excellence by cultivating a spirit of cooperation with its business partners. In the modern complex web of organizations and processes, such a spirit of cooperation has become a necessity. There exist groups of organizations including those in education, healthcare, hospitality, financial services and other service sectors which excel through a well-designed path of cooperation.

Before we briefly explore organizational excellence followed by a comprehensive discussion of the subject in the next chapter, we may have a look at 'operational excellence', which focuses on core and support processes, considers value as some feature of the output (product) that satisfies the customer and for which the customer is ready to pay, identifies value-adding and value-neutral processes, lays stress on the flow of the value stream across processes or activities and ensures that the flow does not break down. Thus operational excellence is concerned with the efficient use of resources guided by an effective strategy. According to Brovkin (2024), "Operational Excellence enables an organization to do more with the same staff through better employee engagement and streamlined processes. It is not about cutting resources, but about figuring out together how we can better apply our resources." Operational excellence is a growth-oriented framework focused on customer satisfaction through employee engagement and lean management principles (not necessarily by naming the concept), not cutting down on resources but cutting down on waste. The framework also emphasizes continuous improvement. Duggan (2016), another expert on operational excellence, refers to several key principles under this framework: (1) design lean value streams; (2) make lean value streams flow; (3) make the flow visual; (4) create standard work for flow; (5) make abnormal flow visual; (6) create standard work for abnormal flow; (7) have employees in the flow improve the flow and (8) perform offense activities. It is obvious that this scheme lays emphasis on efficiency in resource utilization by employees through the use of relatively simple and visually convincing tools. In fact, a number of consulting organizations have put forward a host of strategies to achieve operational excellence, some of which take advantage of problem-solving methodologies like Six Sigma and total productive maintenance to boost productivity.

However, it must be kept in mind that operational excellence is not to be equated to organizational excellence, the latter implying several features of the organizations beyond achieving operational excellence on a sustained basis. 'Doing more' results in 'better' results in terms of measures of organizational performance across different dimensions. It may not be wrong to claim that sustained operational excellence puts the organization on the track towards 'organizational excellence'.

1.5 ORGANIZATIONAL EXCELLENCE

Organizations operate at three distinct levels. Some struggle to survive in a highly competitive environment, characterized by a lot of uncertainty and a remarkable change in attitudes and aptitudes as well as expectations and aspirations of people within and outside the organization. Many of these organizations are likely to close down in the near future, unless a miracle happens. More fortunate are organizations which can sustain their current levels of performance and even are growing, going by some operational definition of 'growth'. These organizations have been trying to standardize their processes in conformity with some national or even international standards and benchmarking their processes as well as results against the best-in-class performers. Only a few organizations are putting up 'excellent' performance currently, over the recent past or even since they started functioning. Also seen are some organizations who were putting up high levels of performance during some period but could not hold on to that level thereafter. And these few have been focusing on continuous improvement and on innovation, resulting in new products and services to capture a larger share in the existing markets and even to create new markets.

A standard or even a widely accepted definition of organizational excellence is difficult to articulate. In fact, while tracing the transition from Total Quality Management through performance management to business excellence, Unnikrishnan (2019) lists 14 definitions of business excellence, including some that differ from a few others only semantically. However, some traits or features of an organization that are associated with the notion of excellence include 'doing things differently' from peers and other constituents of the concerned group. That way, excellence demands something beyond conformity with the latest relevant standards. Excellence also demands a commitment toward continuous improvement as against complacent sustenance of the current level of performance. One can certainly add that if an organization aims at excellence and stays firm on the right track for achieving excellence, it is just likely that it outperforms peers.

Organizational excellence refers to excellence in organizational performance, with the word 'performance' having two clearly discernible connotations: execution and results. Thus, organizational performance would be understood in terms of execution of different operations carried out by the organization as well the results expressed in various dimensions. Quite often, organizational excellence will be referred to as business excellence. Not to do away with this practice, we need a generic concept of 'business' that is free from any stigma or disdain and where 'business' is seen as a necessary support to society. It is worthwhile to mention that such a view of 'business' has been incorporated in the Charter of Establishment of the Harvard Business School where 'business' is interpreted as 'delivering value addition, with profit'. Of the three aspects of 'value'—use or functional value, prestige or esteem value and salvage or re-use value—some or all may apply to any organization, of course with an appropriate contextualization of the words 'value' and 'profit'. Thus almost all the organizations that can be characterized by this extended definition of 'business' can be justifiably brought under the umbrella of 'business' and the concept of business excellence will be in order.

Basically, measurement in the context of business excellence really refers to measurement of business performance. The main focus for commercial organizations is on financial results. However, the better-known business excellence models are primarily derived from quality management models like the Malcolm Baldridge National Quality Award (MBNQA) model, the European Foundation for Quality Management (EFQM) model and similar national models. Evidently, the focus in such models lies on process performance, starting with the standards adopted and reflected in financial results, stakeholder satisfaction, impact on society and impact on the environment. For each aspect of process performance and results, several indices have been suggested over the years. While uniqueness of any such index is too much to expect, certain criteria for any measure of performance have been discussed by researchers, which are covered in chapters 3, 4 and 5.

While linking results with different aspects of process performance is the primary task of building a model to be used for the purpose of initiating improvements to the existing state of affairs, the task of combining measures related to different aspects of 'results' affecting different stakeholder groups into a single yardstick requires assignment of appropriate weights to the component measures and this poses serious problems.

In some analyses of organizational excellence, the performance of an organization in important processes and results thereof are taken as reflections of endogenous factors, mostly related to the management of resources—physical, material, technological and human—by the leaders by adopting some strategy. Some other explanations attach due importance to externalities that affect performance. It may not be always possible to envisage external environmental factors that are likely to adversely affect performance, making performance not completely explainable in terms of assignable factors only. Such 'chance' factors may be associated with changes in macro- or even meta-level socio-politico-economic situations, leaving aside unpredictable changes in the weather.

1.6 ETHICS AND VALUES

The rising phenomenon in which some organizations encourage a value system that may not respect age-old virtues like honesty and integrity and adopts unethical practices to demonstrate high levels of achievement points to the critical role of values and ethics in determining or influencing organizational performance. These values and ethics (or the absence thereof) are reflected in the organization's policies and strategies and even in their visions and missions.

Values are beliefs that a person holds about different aspects of life. These are guiding principles that moulds a person's desires, feelings and actions all through his life. (Values are acquired or formed over years of life in different ways.) To one facing a dilemma, these values guide one's choice of action and behaviour. And thus values are essential determinants of decisions in routine as well as challenging situations faced by an individual. Positive values like honesty, integrity, respect for others, adherence to norms accepted and promoted by the organization and sincerity do contribute to enhanced competence and better performance. Since new values are acquired and some old ones are modified by individuals through learning and experience, it becomes imperative for management to ensure a continued alignment of individual values with organizational values.

Different societies have different value systems and the behaviour and actions of people in a society are guided by these values. People also have their own value systems which act as the lens through which they view the world around them and which provide them a basis to make judgements. This infuses some complexity in situations where a contradiction arises between values cherished by a society and personal values held by an individual. Such conflicts create problems in the society as well as to the individual.

Organizations need to have their respective value systems to guide their workforce to choose between what is 'good' for the organization and what is 'bad'. As expected, leaders have the responsibility of delineating such a value system and communicate the same to all constituents of the organization. Leaders also have to comply with 'business ethics' and even relevant 'professional ethics'. Short-term gains through bending of rules and cutting corners while negotiating have to be discouraged. Leaders must first follow ethical standards and organizational values themselves and then encourage and enable others to practice them. They can draw inspiration from success stories of organization—maybe a few in number—which have scrupulously adhered to relevant ethical standards and pursued their own organizational value systems.

1.7 REVIEW OF LITERATURE

Organizational excellence has been a matter of perception and appreciation, primarily by clients and customers since long. However, it drew the explicit attention of management professionals and researchers who came up with the publication of reports, articles and books on the subject only in the 1980s. In fact, many researchers in the field recognize the book by Peters and Waterman (1982) titled *In Search of Excellence: Lessons from America's Best-run Companies* as the first published material in this field. However, quite a few books and articles on the subject—without an explicit mention of the term 'organizational excellence'—had been guiding the industry and businesses over the last seven or eight decades, particularly emphasizing on quality improvement as was being practiced in Japan as also in some leading American and European countries. Many publications of the Union of Japanese Scientists and Engineers, the American Society for Quality and the European Organization for Quality appeared since 1940s. The concept of Total Quality Management went through discussions to be subsequently recognized as a part of organizational excellence. Also, literature on management, especially on organizational development and focusing on people involvement and use of quantitative analysis to promote organizational performance, has added much to the literature on the subject.

Particular mention should be made to materials published in the journal *Total Quality Management* over the years, with the name of the journal changed to *Total Quality Management and Business Excellence*. During those years, 'business excellence' stood for 'organizational excellence and a host of articles reflecting diverse forms of organizational excellence and carving out different roadmaps for achieving excellence have been appearing in this journal over the last few decades. In fact, this journal included at least one article on this topic in almost every issue. And, as expected, many of these articles highlighted the transition of Total Quality Management to organizational excellence through operational excellence.

Literature on organizational excellence covering a whole range of issues right from the concepts and constructs of organizational excellence, through business excellence models, key performance indicators and critical success factors, roles of leadership and planning, contributions of continuous process improvement and innovation, importance of change and adaptation and concern for the society that supports organizations has grown rapidly, in the process covering different types of organizations. Quite a few journals exclusively devoted to topics like change management, technological innovation, business process management and the like have emerged over the last few decades, with the result that several periodicals contain articles relevant to organizational excellence, besides quite a few journals on management theory and practice. In addition, there are interesting blog posts and articles on the web by consultants, academicians and practitioners. All this makes the task of reviewing even current literature on organizational excellence not just onerous, but possibly not that meaningful. It has been felt that literature on any aspect of organizational excellence should better be attempted in the chapter that deals with that particular aspect, which is what has been done in this chapter.

Some recent reviews make the task of literature review a bit simpler in some sense, though a few of them eventually end up in what can be referred to as a GCM (greatest common divisor or highest common factor) of the materials covered. Mention may be made of Ubaid et al. (2020), Bernawi (2022) and Ranjibarian (2022), among quite a few others. Of course, there are some contributions like *Good to Great: Why Some Companies Make the Leap and Others Don't* by Jim Collins focusing on organizational transformation under a disciplined leadership, *Leading Change* by John Kotter emphasizing management of change as a key factor to be considered and *And Suddenly Appears the Innovator* by Genrich Altshuller proposing his theory of inventive problem-solving for new product development.

In the domain of measuring organizational performance in the journey towards excellence, one can hardly miss the milestone created by the balanced (business) score card introduced by Kaplan and Norton (1992). Drawing upon the need for a comprehensive view of performance by strategic management to diagnose its current state and to identify areas of potential improvement, this score card provides a few important measures of performance from each of four perspectives: stakeholders, customer, internal business processes and capacity for learning and innovation. Quite interestingly, the concept of triple bottom line (Elkington) comprising people, planet and profit reflected a worldwide concern for sustainability and a call to business to stress on the persistence of organizational performance in terms of both financial and non-financial metrics. Similarly important is the concept of corporate social performance, which includes corporate social responsibility.

Several models to explain business excellence including the European Foundation for Quality Management, the Canada Excellence Model and the IberoAmerian model, besides the models indirectly involved in the famous quality awards like the Malcolm Baldridge National Quality Award, the Deming Prize and similar awards have been discussed by a host of authors, some of them using structural equation modelling to explain the theory behind the model, some focusing on the extent to which the 'drivers' of excellence, such as leadership, strategy and policy, market focus, resource management and process management, are currently being harnessed and their contributions to the 'results' on different fronts being currently achieved. An excellent review of business excellence models both providing a framework to guide management in moving towards excellence as well as for external assessment leading possibly to an appreciation or award has been presented by Dahlgaard et al. (2013). The authors point out the limitations of these models in predicting future performance and therefore offer a simplified 4-P model. Leadership as a concept and its characteristics as well as its crucial role in navigating organizations through the increasingly rough and disruptive environment has been a major topic for discussions in the broad area of management and especially in the area of organization development. Among the many theories of leadership, Covey's classic model of seven habits of effective leaders, encompassing personal traits, inter-personal abilities and self-renewal, stands out as a pace-setter. Looking at the list of roles expected to be played by a leader to navigate the organization through fair as well as turbulent weather, we find a whole range starting from an effective communicator, a good strategist, a guide and a supervisor, to a risk manager. A simple model that appeals to the author as quite actionable is the 4-E model where leaders are expected to envision (develop and disseminate a vision to be shared by all), encourage (people to accept and welcome the vision), enable (of enhancing knowledge and skills) and empower (to apply requisite knowledge and skill without any hindrance in discharging responsibilities) the entire workforce. It would be wrong to claim though that this is the model to be adopted in each case.

Leadership at all levels has to plan, weighing different options on different counts (likely impacts on business) to arrive at some decision meant for implementation. For corporate leadership to develop an organization-wide long-term strategy, scenario planning and analysis has been a widely discussed subject. Works by Konno et al.(2014) are quite worth mentioning.

Process management plays a crucial role in enhancing operational efficiency and putting the organization on the right track to excellence. Right from the simple tools for process analysis and improvement started in post-war Japan like kaizen, 5-S, judoka, kanban, poka yoke to methodologies like Six Sigma, eight-D or lean management or total productive maintenance for working out breakthrough improvements to the deployment of Artificial Intelligence-enhanced business process management, a whole host of methods and techniques has been elaborated by many authors and institutions.

Without innovation, no organization in the current business environment can perform satisfactorily on a sustained basis and this calls for the institutionalization of innovation and creative thinking to come up with innovative products and services. The theory of inventive problem-solving (TRIZ) by Altshuller and its software-based version ARIZ provides details for new product development and DeBono discusses creativity in concrete terms. Incidentally, a journal *TRIZ* is devoted to innovation and innovative problem-solving exercises propounded by Altshuller. Not all innovations involve the development or adaptation of new processes and/or new technologies. With the rapid pace of technological advances, any organization not developing a new technology, not upgrading its existing technology or embracing a new technology will be soon left behind its peers who take due advantage of technological innovation beyond digital transformation, which has come to stay in most organizations.

All said and done, organizations intending to stay on the path to excellence have to revisit its culture and climate, its business model and structure, goals and objectives, formal processes and performance profiles. Thus change management becomes a great challenge since change initiatives are met with resistance to some extent. The subject has become so important that an international journal is being published exclusively devoted to change management, and a few other periodicals have also published many original and review articles regularly on the subject.

The importance of values and ethics and the need for developing an organizational culture that pervades the organization and instills a sense of cohesion and team work among employees based on some ethical principles of behaviour has been highlighted by many authors. Hultman and his associates have extensively deliberated on issues linked to the adherence to ethics and values, on the one hand, and skills and aptitude, on the other, in building up organizational competence. Hamid (2015) speaks about value-based performance excellence for higher educational institutions, emphasizing both tangible and intangible aspects of performance. In this case, an important, though not easily obtainable aspect of performance, in terms of the roles and responsibilities along with achievements eventually secured by the host of students who leave the portals of the institutions to join the world of work, usually remains outside the boundary of a performance metric or a score card.

Significant advances in technology for the production of goods and services besides those in the domain of digital technology influencing many different processes planned and executed by organizations of various types have taken place at a fast pace over the last few decades. Leaving aside specific technologies of manufacturing, digital technologies that offer support to such technologies, through devices and equipment assisted by artificial intelligence, which results in more efficient production of better goods and services, have been absorbed by many forward-looking organizations. Organizations in the fields of education and research have taken great strides in this direction. A review of the three distinct levels of applications recognized commonly as digitization (really implying automation of processes), digitalization (taking due advantage of distributed data storage and processing, machine learning, high-speed computing etc. to improve processes radically) and digital transformation (leading to a changed business model altogether with a change in product and service offerings) would be a stupendous task and interested professionals and academics can seek the relevant literature including case studies and descriptions of methods. What has been stressed in this book is the fact that responsible use of artificial intelligence and related developments in digital technology is fast becoming a necessity for survival and not just a golden opportunity in the journey towards excellence by a few forward-looking organizations.

Given the virtual deluge of literature on the subject of organizational excellence directly and on related topics in management, organization development, leadership, business process management, innovation, human psychology and corporate performance, one is quite likely to confront divergent views sometimes irreconcilable, different approaches, distinctive perspectives, varying stress on quantification and use of quantitative methods and tools, and even references to uncommon phenomena treated in other branches of knowledge like evolutionary biology or social networking. This is what should make a conscientious reader think more about excellence, learn more about its ramifications and contribute more to the subject through thoughts and expressions.

1.8 SCOPE OF THE DISCUSSION

The present volume is not intended to be a textbook that can be readily prescribed to students attending a formal course on management, or on organizational development. It can modestly claim the status of a reference book for serious students and researchers as well as professionals interested in the subject, which is of great contemporary interest.

The discussions in the present volume centre around a somewhat broad and a bit unconventional assortment of organizations, which include academic and research establishments. The author attempts to characterize 'organizational excellence' as a journey rather than an end and distinguishes it from correlates like growth, success or competitive advantage. Excellence in performance as reflected in the way an organization functions and the results that emerge has been discussed from different perspectives. Given the extensive publications on the subject, the focus has been on not-much-discussed concepts, methods and techniques, including some, which may appear to be a bit odd. To enable a self-complete volume, traditional as well as more recent contributions have also been covered, occasionally with some critical observations by the author. These materials are scattered in chapters 5, 8, 9, 10 and 11

Speaking about measures of organizational performance, the triple bottom line and the balanced score card have been dwelt upon, with a small discussion on the extended performance satisfaction index. A section on the more recent topic, viz. corporate social performance model, has also been included. While reviewing the more established business excellence models like MBNQA or EFQM, the Australian Business Excellence Framework and the Brazilian Management Excellence Model MEG along with the Charter Mark for the public service sector find a place here. Among the drivers of excellence, leadership and strategy have been explained in relevant details. Some ideas about servant leadership and, in the last chapter, value-based leadership have been included. The distinctive features of organizational decision-making under uncertainty as compared to traditional decision-making based on expected pay-off (utility) have been presented and some models for business decision-making have been discussed. Scenario analysis with some original contribution of the author has been offered as a critical input into strategy development, along with robust decision-making.

While accepting the crucial roles of leadership and strategy as well as organizational culture and organizational climate in influencing organizational performance, the journey towards excellence does emphasize effective and efficient management of core and support business processes. A brief summary of business process re-engineering and of various process improvement approaches and methods has been provided. The book also focusses on innovation and new product development, providing an exposure to TRIZ (theory of inventive problem-solving). Acknowledging the fact that not all innovations are technological in nature and that new technologies emerging at a rapid face will change the very architecture of all types of organizations, a chapter has been devoted to a discussion of technology development as well as on technology adoption as requisite steps in the long march towards excellence. The strong need to initiate, steer and manage changes that are necessary in all aspects and at all times in all activities of any organization in the face of some anticipated and some unforeseen hindrances has been addressed in the penultimate chapter. In this connection, simplistic diagrammatic tools used in the theory of constraints have been incorporated.

With a firm belief that an organization committed to the cause of excellence and to sustaining a reasonably high level of performance has to be necessarily led by a corporate leadership that is aligned to the core values reflected in the vision and mission of the organization, which is practiced by the entire workforce, the book concludes by expounding the essentials of ethics and values, their adoption and practice as well as good governance coupled with activities undertaken as part of corporate social responsibility to be aligned with sustainable development goals or similar other mandates to benefit humanity.

The central theme presented in the pages to follow is that the quest for excellence has to be guided by facts, taking due advantage of data analytics to measure different aspects of performance, embracing advances in technology with due adaptation to derive beneficial effects only, developing and sustaining a conducive workplace climate and a culture woven around the vision and mission of the organization and steering through ups and downs in business by a visionary and committed leadership, which absorbs and practices the core values and ethical principles embodied in the charter of the organization.

The content includes a comprehensive discussion on measuring organizational performance not merely for traditional business organizations but also for academic institutions and R & D establishments. Business excellence models have been briefly but critically reviewed and some problems encountered when using any of these models are also covered. In the journey to excellence, the author has considered the roles of change management along with the existing improvement methodologies. An attempt has been made to indicate the role of innovation in this context and some broad guidelines for institutionalization of innovation have been provided. The overarching role of continuous improvement using contemporaneous developments in mapping, analysis and optimization of business processes has been highlighted.

However, the limitations of this book also have to be informed. Unlike other volumes on management in general and organization development in particular, the present volume does not contain any case studies, though indicative illustrations have been provided to some extent. These are in reference to anonymous organizations in several different domains. This can be explained by the view taken by the author that 'organizational excellence' is an endless journey and not a fixed goal where this journey ends. This view leads to the author's reluctance to accept any organization, even with a remarkable all-round performance sustained over a period till date, as 'excellent'. Components of performance or their determinants in some 'great' establishments have, however, been referred to in some places.

The review of extant literature on the subject cannot be taken as even near complete, and the intention is only to put forth some ideas that seemed to be useful and interesting logically. Needless to say some of the ideas could have been already written or otherwise spelt out by other exponents of organizational excellence, beyond the limited knowledge of the author.

2 Organizational Excellence

2.1 A JOURNEY, NOT AN END

An attempt to develop a concrete and comprehensive idea about organizational excellence or even an operational definition of the term that can convey the innate meaning and can provide a framework for actions may be based on the common themes in the definitions proposed by eminent people. One problem that arises in this process is the inconsistency—maybe, apparent to this author—in some of the expositions. While a 'journey' can be comfortably taken as implying the same innate concept as 'a continuing effort', it may be difficult to place 'a state of accomplishment' in the same category. Similarly, the use of the phrase 'can be achieved' is somewhat difficult to reconcile with a statement that 'organizational excellence is an ongoing effort'. If we like to refer to organizational excellence as a goal, we should qualify the term 'goal' as 'receding' and not as 'pre-fixed'. Otherwise, the focus on 'continuity' of the efforts or the journey will be lost.

It is commonly agreed that excellent organizations do not just conform to certain standards relevant to their various business processes; instead, they attempt to do better. Even then, it may not be inappropriate to begin our discussion with a definition or explanation of organizational excellence offered by the American Society for Quality (ASQ), since excellence is conceptually related to the quality of performance and ASQ standards for various facets of quality have been found quite useful by a wide range of organizations. ASQ considers organizational excellence not as an end, but as a means to the end. The word 'end' may imply a goal, which arises from a vision about the future of the organization. The ASQ proposes the following definition:

> Organizational excellence is defined as the ongoing efforts to establish an internal framework of standards and processes intended to engage and motivate employees to deliver products and services that fulfill customer requirements within business expectations. It is an achievement by an organization of consistent superior performance—for example, outputs that exceed meeting objectives, needs or expectations.

Two statements in this definition make it amply clear that organizational excellence implies 'ongoing' efforts and 'consistent superior performance' through an 'internal'

DOI: 10.1201/9781003591733-2

framework of standards and processes meeting such standards. Thus, organizational excellence does not refer to a terminal or a now-and-here performance that is superior to that of peers and competitors. For a monopoly, excellence means a sustained improvement in performance (measured by suitable indicators) over the years till date, at least for several years beyond a certain transition point defined by the organization, or right from the very beginning. Thus, this definition may not strictly apply to (recent) start-ups. The efforts to improve within a framework developed internally must be continuing, measured through performance indicators. This framework eventually corresponds to the framework for policy and strategy.

ASQ goes on to define excellence as a "measure of consistently superior performance that surpasses requirements and expectations without demonstrating significant flaws or waste." This statement implicitly refers to degrees of excellence that can be achieved, which is rather avoided in our discussion.

In one of the most-read books on the subject titled *Good to Great*, the author Jim Collins avoids a direct use of the phrase 'organizational excellence', though 'greatness' or becoming 'lastingly great' can be thought of as implying 'excellence'. He refers to a sustained superior performance (over several consecutive years following a transition point)—primarily in terms of financial results—as defining the transition from 'good' to 'great'. Of course, his classification of 'great' organizations does cover all the important decisions and actions that an organization has to take in order to stay on the path to become 'lastingly great'.

A consultant Farghaly (2023) defines organizational excellence as a state of remarkable performance in all aspects of an organization, including its culture, leadership, people, processes and results. It is a state of continuous improvement where the organization strives for excellence in everything it does, and where every member of the organization is committed to achieving excellence to satisfy and exceed stakeholder expectations. He goes on to elaborate each of the seven key areas that deserve the attention of any organization. This definition considers 'excellence' as a 'state', without an explicit mention of this 'state of remarkable performance' as something too difficult to reach and which has to be maintained over a reasonably long period till date, not allowing a breather now. Another problem is to concretize the adjective 'remarkable'. Several peer organizations may have been showing 'remarkable' performance. Taking 'remarkable' in the context of each organization separately, no organization possibly outperforms others. Performance in terms of 'results' is easier to assess compared to performance in the other six aspects. The author refers to the aspect 'culture' as the one that encourages innovation, continuous improvement and pursuit of excellence in all its activities.

The characterization of organizational excellence as a journey (without an end), as we like to pursue in this volume, is bound to be a bit ornamental. We state that organizational excellence is characterized by a firm determination to undertake a long and arduous journey towards a remote and receding peak, following an uneven and slippery path with unanticipated potholes (which can entrap you and hold you back, at least for a while). The path to the peak becomes steeper as you move up further, through rough weather with occasional moments of relief and rejoice, but the organization is motivated by a clear view of the peak and a feeling of confidence to reach that peak at some time. And the journey, like a pilgrim group, has many a trekker with diverse interests and skills but unified by the ardent desire to reach the peak and be rewarded for that. The group has to remain aware of other groups proceeding along the same path, possibly with the same ultimate objective, trying to nudge or shove the group out of the way. The group has to use tools that help it stay erect and move forward despite such hindrances.

We can, and possibly should, wind up this section and, instead of noting the deficiencies in different definitions and problems in accepting them, we now proceed to characterize 'organizational excellence' in terms of both latent and manifest features, with a focus on those aspects that allow an objective, if not unequivocal, assessment.

2.2 CHARACTERISTICS OF ORGANIZATIONAL EXCELLENCE

Having accepted the view that organizational excellence implies a continuing effort without a breather, encouraged by the vision and the mission and not undertaken following a threat or a challenge, we can pointedly look for some features of an organization that is set on the right journey towards excellence. Some of these features will be quite visible and thus can provide some guidance to organizations intending to take up this journey, while there will be several features incorporated in the way leaders behave, people work, processes are carried out, procedures are followed, assessments are made, technological innovation takes place, societal concerns are addressed and in many other aspects of the organization's functioning. In the context of process benchmarking as an exercise by an intending organization, such latent features maybe addressed, if feasible. Many management science experts have provided elaborate characterizations of organizational excellence and we will consider some of them.

Keeping aside semantic problems, it should be noted that quite a few of the 'characteristics' of organizational excellence can be and are looked upon as the 'determinants' or 'drivers' of excellence, while a few others are really the 'results' of excellence. Generally speaking, most results are manifest, while more of the determinants need a closer look and are not so externally manifest. Further, the features or traits of organizational excellence should not be identified with traits to be possessed and displayed by individuals (e.g. leaders or managers) within the organization, and should relate, instead, to the collective behaviour of the entire workforce. Some of the features of organizational excellence can be explained in terms of some theory concerning organic or other types of systems. The characteristics of organizations steadfastly proceeding towards excellence have been enumerated and explained by many distinguished authors. The author, however, likes to present the features discussed by a lesser known expert Robbertze (2021), with minimal editing and modification, to start with.

Robbertze considers four sets of features: (1) noteworthy performance; (2) happy employees; (3) managed resilience; and (4) social responsibility (responsibility towards society). The first group comprises four traits—mostly visible—relating to (a) fulfilled purpose, (b) financial stability and independence, (c) continuous growth and (d) customer satisfaction. It may be pointed out that the purpose spelt out initially and followed consistently thereafter should reflect an acclaimed value system that would attract investors, workers and customers. The fulfilled purpose—even if the purpose is a modest one—sums up quite a few decisions and actions taken by the organization. The organization should have generated enough funds through marketing of its products and services to meet its ongoing operating expenses and also to ensure the availability of funds needed for both planned operational change and emergency funds for adverse economic situations. Continuous growth in terms of market expansion and expansion of product and service profiles, possibly to take care of innovation and new product service development, has been included. These three features do lead to customer satisfaction, as can be gauged from the retention rate and the net promotion score. The performance of processes and results thereof are to be noteworthy, not necessarily superior to that of peers and competitors. Employees must find the workplace environment quite conducive to productive work, beyond meeting the physical needs of convenience, safety and security. A leading feature to enhance people engagement is role clarity, stating the responsibilities for each worker along with commensurate authority or control over one's actions. Encouraging and enabling people to come up with innovative ideas about their respective tasks and also those of others contribute to happiness of the workforce, especially when some of their ideas are accepted and implemented. Collaboration among all the workers through an alignment of individual aims and values in life with the organization's goals and values—not an easy task—has to be accomplished. Workers are to be supported by way of adequate resources they need as well as opportunities to enhance their capabilities. In fact, honest and visible efforts to enable workers to realize their full potential, making a fair assessment of their contributions and recognizing them through awards and appreciation whenever deemed fit make the organizations striving to achieve excellence stand out from the crowd. Then comes the ability of the organization to accommodate and adapt to evolutionary changes, which are a result of customer choices or societal responses and reactions. Also important is the ability showcased by the organization in recovering from a threatening or challenging situation facing the particular organization or even the relevant group of organizations as a whole. This might have been possible through changes in business models and/or restructuring of investment patterns or reduction in debts. In the last category of features, we have the much-discussed corporate social responsibility (CSR) as reflecting the commitment of an organization to extend some benefits to society in return for the direct and indirect support provided by the society without which organizations could not function. And, beyond CSR, which can even be directed towards environmental protection and enrichment, Robbertze points to enabling cyclical economy to minimize the consumption of non-renewable resources and reuse and recycling of used products.

2.3 GROWTH VS. EXCELLENCE

Growth and excellence are closely related concepts and are manifested together in many situations. However, not too often, we come across cases where growth is manifest but excellence is nowhere visible. Similarly, there are cases where organizations have not grown much but excel compared to others (including some which have been growing) in terms of performance. Thus, the relationship between growth and excellence has been somewhat debatable and management professionals and researchers are divided in their views on the issue. The following statements reveal this divergence of views to some extent. (1) Growth and excellence can both be achieved simultaneously. (2) Excellence does not necessarily imply growth. Too much of growth may lead to erosion of excellence. (3) To achieve excellence calls for a growth mind-set. Without growth, the question of excellence may not be relevant.

Sometimes these two concepts have been held as mutually contradictory, raking up the quantity–quality controversy. As a rather unfortunate instance is the fact that a service delivery organization in the healthcare education sector marked by excellent organizational performance may witness a growth in numbers of potential beneficiaries without a matching augmentation in resources and may end up losing the 'excellence' rank in the assessment of the tasks performed or the results obtained.

In the fitness of things, a concrete differentiation—which may not apply universally or is to be accepted as 'reasonable' at the least—should be attempted. Towards this, let us recognize the following three entities as distinguishable quite objectively:

1. Inputs to processes in terms of physical, material, technological and human resources acquired and deployed
2. Processes performed and procedures followed to carry out activities within each process and
3. Results achieved in different fronts or along different dimensions of results impacting different stakeholder groups.

That growth does not necessarily imply excellence and that even 'planned' growth may hinder the pursuit of excellence has been evident in the performance trajectories of many organizations in different sectors. Glaring examples of the growth of academic institutions resulting in a decline in the quality of performance are on the increase. Several schools, colleges and universities started and managed by private groups have been registering growth in terms of the number of students, faculty members and other supporting staff (though less than proportionately), number of courses being offered, facilities for extra-curricular activities and growth of independent thinking, cooperation and collaboration with other institutions—domestic and foreign, projects awarded by different funding agencies, higher ranks in some national or international ranking frameworks and sometimes the lucrative job offers and other achievements of a few promising students or those graduating from the institution. At the same time, the level of knowledge and skill acquired by the majority of students and the quality of their performance in work life have been unimpressive. In fact, whatever they could justifiably claim to have achieved when they were small and were focusing more on their distinctiveness from the existing peers has been gradually lost as the emphasis shifted from quality to growth. In fact, their initial focus on quality led to their expansion and growth—maybe, beyond a desired level—and led to the subsequent deterioration in standards.

One should not forget the fact that a few forward-looking organizations grew substantially, without lowering to any extent their intents and actions in the march towards excellence. In fact, organizations involved in space research and related development activities including design, development and deployment of launch vehicles and even space ships etc.—mostly in the public sector of some advanced countries—were small organizations doing challenging and innovative work excellently and have grown over the years, remaining firmly on the track to excellence. Indeed, these are some 'great' organizations where growth and excellence have been sustained together.

With regard to manufacturing or service organizations, where excellence or, rather, excellent performance has been noted primarily in terms of financial performance, we generally find many organizations outperforming competitors or peers over a certain time horizon unable to retain their positions after some time. As mentioned by Jim Collins in *Good to Great*, 11 American organizations were rated as 'great' in terms of their performance, while [CE : among] ten others four failed to exhibit the 'good-to-growth' phenomenon and six organizations could not sustain their positions as 'great'. Of course, Collins does not relate the transition from 'good to great' to any growth pattern.

Accepting this differentiation, one can argue that growth is oriented to expansion in resources and results, leading to superior performance in growing organizations. Initially, their superiority is visible in the first entity as well as some aspects of the third, while excellence is more concerned with the second entity as well as some other aspects of results like employee satisfaction or impact on society or on the physical environment. Growth, by and large, lends itself to cardinal measures, while excellence generally requires an ordinal assessment. Growth is more tangible than excellence. Growth is reflected through an increase in workforce, line or verticals of business, physical assets and material stocks, market shares, profit after tax, dividends to shareholders and similar other concrete outcomes. Excellence in performance is surely associated with improved quality of service and value to customers (internal as well as external), business partners, investors and custodians of public interest. It may not necessarily exhibit some of the manifestations of growth.

It may not be out of place to mention right here that Deming—the pioneer of quality management—spoke of the situation where workers are constantly reminded of production targets as a dreaded disease. And if meeting specified targets is regarded as a success, then success and excellence may not go hand in hand.

Growth in resources may be a pre-requisite for excellence, as in the case of start-ups. A threshold level of resource inputs, especially by way of skilled and knowledgeable workers and recent applicable technologies along with a minimum financial capita, is in some cases a necessity for excellence. In a slightly different tune, we may even hint at the need for growth in leadership qualities among workers at all levels of responsibility and authority.

2.4 OTHER CORRELATES

Two other concepts akin to excellence are worth a brief discussion. These are 'success' and 'perfection' along with their implications and scope of application. There is no unanimity in views on any such concept or construct and one should internalize the divergent views appropriately to form their own. And the view to be adopted should lead to an understandable differentiation among the related concepts with a distinction between the overlaps. Let us take up the concepts of success and excellence first.

We may agree on the view that 'success' relates usually to a particular task assigned to a specified segment of the organization to achieve a given objective or target that is likely to result in improved performance by the organization, while excellence commonly applies to the entire organization and all its activities and achievements, without reference to some fixed targets or objectives. Bagging a lucrative contract, modifying and eventually adapting an imported technology to the need and situation of the organization, making the maiden entry into the international market, getting the quality management system duly certified for conformity to the relevant ISO standard, coming up with a new market-ready product or service and the like are well-recognized as successes. These will surely lead to improved organizational performance, at least for some time beyond the tasks being completed. Similarly, starting with a modest or moderate set of goals and objectives that accommodate current faultlines in the organization, it may be possible to achieve the targeted goals and objectives and this achievement could be treated as a success. While these may well be taken as mileposts in the journey of the organization towards excellence, it is also possible that these isolated events might push an organization to complacency and the initial successes are not matched up with continuing efforts to move farther ahead. Doubts may remain about the sustenance of such a success in ensuring sustained improvement in organizational performance. Further, success in one task or process may sometimes be achieved at the cost of a less than satisfactory performance in some other task(s) and may not lead to any improvement in the overall performance of the organization.

Excellence, on the other hand, is oriented towards a commitment to improve performance in all segments of the organizations in all their tasks at all times by all the people concerned. Excellence does not refer to a fixed goal or target.

One way to link success with excellence is to look upon excellence as the eventual outcome of a sequence of successes over time, with success related to some specified target for organizational performance (in terms of a few critical parameters). In some sense, this implies continuous improvement in performance, where in each step we develop and implement a strategy that is geared to achieving a target for that step. That way, the continuous improvement exercise may be focused on stepwise strategies. Somewhat like 'success', the concept of 'perfection' takes care of a fixed goal or target and implies complete achievement of the former, without entailing any waste or redundancies. Perfection characterizes a situation wherein needs and expectations of all stakeholders (assumed to remain at their currently prevailing or anticipated levels) are fully met within business conditions because of the best efforts put up by the entire organization. Generally, perfection is not an attribute of a given segment of the organization or for a given task. Perfection gives the impression of best achievable and thereby indirectly rules out any further improvement effort.

To sum up, we notice both commonalities as well as differentiation among the concepts mentioned earlier: excellence, growth, success and perfection. Improvement in efforts as well as in results creates a common ground, while a fixed versus receding goals and targets, whole or part of the organization getting involved, terminal versus continuing effort and the like tend to differentiate one concept from another. It will be wrong to say that all these concepts are aligned to organizational excellence and a hard differentiation is not warranted in a discourse on organizational excellence, except to bring into focus the essential features of organizational excellence.

2.5 ORGANIZATIONAL RESILIENCE

The long march to excellence is not likely or expected to be a smooth and untested one. In the context of contemporary turbulence in the business environment and the consequent variations in performance, any organization, irrespective of the type of activities being carried out by it, can hardly escape situations of despair during the course of its operations. In some adverse situations caused by extremely unfavourable environmental factors, an organization may receive a rude shock or a potential threat to its very survival. Such crises have to be overcome—as and when these arise—if the organization resolves not merely to survive the shock but to restore its earlier performance level as early as possible thereafter. What is needed is an ability to face the crisis silently, though with pain and temporary setback, and engage in efforts to reach the pre-crisis situation as early as possible without any planned reduction in resources, especially human resources, unless compelled to do so. This ability is generally characterized as resilience and organizational resilience is a pre-requisite for any organization to stay on the path to excellence. Resilience as a virtue or strength of an organization is built up through a conducive far-sighted management philosophy driven by looking beyond the present, remaining prepared for the worst-scenario and tiding over any crisis within a reasonable time.

Organizational resilience can be formally described as the ability of an organization to anticipate and accept, respond to and tide over uncertainties, threats and disruptions. Uncertainties about the supply chain and the market behaviour, threats from complex and unexpected geopolitical situations and disruptions caused by the emergence of new technologies tending to throw out existing products and services from the market illustrate situations that can catch organizations unawares and create severely negative results. Unless resilience has been a part of the organization's culture, continued existence—not to speak of growth and satisfactory performance—may be at stake. Resilience is not synonymous with crisis management, which relates to actions taken to get over a crisis once it has taken place. Crisis management is a response and organizational resilience is a feature of the organization cultivated through some conscious efforts continuously that guide this response. Resilience enables effective crisis management.

Resilience has been interpreted by some researchers as the ability to resume, rebound, bounce back and positively adjust in the face of untoward events. Some take it as an ability to effectively absorb, develop situation-specific responses to and ultimately engage in transformative activities to capitalize on disruptive signals that potentially threaten organizational survival. It is generally appreciated that resilience is not any of the following and may draw upon all of them as required: agility, adaptability, flexibility, improvisation, redundancy and robustness.

Bell (2002) suggests five principles to build up organizational resilience: (1) flexibility of working hours; (2) staff re-allocation, which requires multi-skilling of the staff; (3) minimization of absenteeism and turnover; (4) job re-design and (5) reorganization of operations. Obviously, management has to adopt these principles and put them into operational practice, engaging in efforts to lead the changes involved in the existing work practices by learning from past experiences. Another model to infuse resilience (known as the Cranfield methodology) speaks of four S's to guide actions in response and create disruptions and opportunities. These are (i) the foresight to anticipate, predict and prepare for the future; (ii) insight to interpret and respond to present conditions; (iii) oversight to monitor and review what has happened and to assess the changes; and (iv) hindsight to learn from past experiences.

To study the nature and extent of resilience in any organization, we have to derive some insight into

a. management philosophy in general;
b. management philosophy to tide over crises; and
c. management experience in handling extremely adverse situations along with the consequences suffered.

Top management has to build a culture of support for the entire workforce, encouraging and extending help in some form or the other whenever a member is in need of support to overcome some adverse situation at the workplace, in the family or in the society. This support culture contributes much towards fostering a united action taken by all in the organization in the face of any impending adverse situation developing externally or a similar situation impacting current business prospects. Meaningful communication among people in different functional areas on a regular basis and interactions through meetings, seminars and events to observe some milestones in the journey of the organization could help people understand what should they do in case of a crisis, which can foster a spirit of resilience. Presentations could be made by way of narratives or even stories created to show how fortunes of some renowned organizations in diverse fields have been badly affected by unexpected threats and disruptions that the organizations could not anticipate and could not tide over in relatively short time, eventually leading to disasters. In the same vein, success stories of how some other organizations could build up resilience as an innate strength to withstand the adversities emanating from unknown and unexpected quarters. Even occasional 'drills' to absorb shocks from a critical situation, going beyond physical displays and accommodating suggestions to overcome a shock as quickly as possible may be practiced. Possible shock-absorbing suggestions could be changing the product profile to come up with new products that can be conveniently produced by the production facilities available, only changes in inputs being required or/and cutting down administrative and overhead costs by changing workplace and time schedules, or/and renting out certain facilities and resources like calibration of equipment to other interested parties, making use of slack employee times to maintain internal assets and so on. To enable workers on certain specific jobs being currently handled by them to be put on different jobs in case of a crisis, multi-skilling of workers by training them in several different jobs has to be promoted by the management.

One way to obtain the desired insights will be to develop suitable questionnaires to be shared among decision-making and executive management personnel, and some selected workers at operational levels should also be included in this exercise. Evidently, we should include more questions on segments (b) and (c) above.

While open-ended questions seeking responses from the respondents not necessarily in the pre-specified categories but capturing responses as these are provided is a better way to gain insights; closed-form questions with pre-specified response categories are easier to analyse. Of course, a mix of these two types also can be accommodated in an enquiry. Framing of questions and specifying response categories cannot be uniquely done. Below are listed several questions intended to elicit the perceptions of the workforce on the 'resilience' aspect of their organization. The list is generic and indicative, through specificities of organizations should be taken care of in the framing of the questions and designing the response categories from which respondents will make their choices.

1. Whether some specialized teams were formed to handle the situation? If yes, what kind of role was assigned to them and what kind of autonomy and empowerment was given to them?
2. Whether all the parts of organization were having similar strain or some parts were having more strain? How this was handled, that is, how more vulnerable parts were given support or resources?
3. The COVID-19 pandemic unfolded in multiple waves and crises were coming up one after the other, leaving not enough time to prepare for the next wave. Was leadership changed in these circumstances and how?
4. Were some steps taken or changes made to enable the organization to handle the series of crises?
5. Was it felt necessary to motivate the people to generate hope as many people have lost their near and dear ones during the pandemic? If yes, how was it done?
6. Whether pandemic crises could be considered as consisting of different phases in respect to efforts taken to successful handling of crisis by your organization?
7. What steps were taken to arrest the spread of shock waves or panic spells across different segments of stakeholders?
8. What was done to ensure quick recovery from the shock and its after-effects? How did these steps help?
9. Were unit heads or in-charges engaged in a discussion with senior management and provided suggestions to tide over the crisis?
10. Even the darkest cloud has a silver lining. Could something be learned from the ongoing crisis?
11. None gains if the organization suffers badly. Did this feeling help in improving the coordination and collaboration during the crisis?
12. Everyone should gain if the organization can steer clear of a current crisis. Was this realization working behind the success?
13. It is said that thinking out of box and putting innovative ideas into practice tones up organization's capability to face the disruptive situation. Were some innovative ideas received and implemented to come out with innovative solutions?
14. Should a free and frank expression about the organization's strengths and weaknesses be encouraged to handle difficult situations?
15. Do you think that employees' involvement in developing an organization's mission and targets contributes to the achievement of these targets?

16. Did this organization face a severe and prolonged crisis anytime within your period of service? If so, when did it take place?
17. Was the crisis you can recall best due to external factors associated with vendors, customers, financiers or regulatory bodies? Or, was it due to internal problems involving employees?
18. What was the role played by regular and contract workers and supervisors? Enumerate briefly the steps taken by the management to overcome the crisis. What, according to you, was the most effective step?
    State the impact of the crisis in terms of loss in revenues, loss of customers and exit of knowledge workers.
19. How much time did the organization take to come back to its pre-crisis situation?
20. Do you want to tell some additional important point, which can be considered for successfully handling the crisis that resulted from the COVID pandemic?
21. Please share your position in the organization (optional).

Questions may also be included to seek views of employees on the role of executive management and to what extent they agree with statements like the following:

Executive management must share apprehensions and realities about impending adverse situations in time and with

(a) top management; (b) business partners and senior employees

Executive management must seek suggestions from each of the above categories of individuals / institutions to tide over the crisis.

Executive management must enlist explicit support from each of the above categories in overcoming the crisis.

Executive management should try to ensure quick recovery from the shock and its after-effects. Executive management must take steps to arrest spread off shock waves or panic spells across different segments of business.

Bearing on the first aspect of resilience, the following suggestions may be considered.

Thinking out of box and putting innovative ideas into practice tones up organization's capability.

Flagging problems is not enough; you have to suggest implementable solutions.

Free and frank expressions about the organization's strengths and weaknesses should be encouraged.

Employees' involvement in developing an organization's mission and targets contributes to the achievement of these targets.

Productivity is not just about producing enough.

2.6 JOINING THE MARCH TOWARDS EXCELLENCE

To join the march towards excellence, any organization at any time since its formation has to commit itself to the culture of continuous improvement, setting high and yet achievable goals, cultivating leadership to develop leaders among people, fostering a spirit of cooperation and collaboration among business partners, encouraging clear and informative communication internally as well as with the external world, promoting free and innovative thinking, caring for sustainability of the resources being deployed in its various processes and conforming to the regulatory frameworks that may be in place from time to time. The list may appear to be rather long. However, quite a few of these 'pre-conditions' go hand in hand with one another and an honest desire coupled with an earnest effort can well sum up the entire story. In fact, the top leadership of the organization has to recognize that "we must not stay as we are, doing always what was done last time, or we will stick in the mud" (George Bernard Shaw). Getting stuck in the mud is a great threat even to the survival of the organization.

To kick-start the journey to excellence, the desire to perform business processes better and more effectively and to produce better results has to be matched with a readiness to do all that is necessary to accomplish the desired improvement goals. The author interacted with the organizational excellence groups initially formed in two different organizations—one in steel-making and the other in banking services—both in the private sector already showing impressive results. Each group was a small team of young, committed senior executives with little traditional staff support, fully authorized to interact with external experts and to carry out small-scale experiments with new ideas, procedures and processes. With no pressure of routine production or sales targets, the group was charged with the task of coming up within a reasonable time frame a few improvement measures that could be incorporated into the existing system. The author found that most of the time initially, members of the group were learning and discussing new ideas, thoughts, methods, techniques and practices threadbare to identify possible ways to adopt the new regime to the existing system.

Left to itself, the group created for steering the organization on its journey to excellence will surely identify quite a few areas in the existing system where improvements can be worked out. However, resource considerations may constrain the choice of areas for improvement to only those where improvement exercises can produce visible and impressive results without much of resource augmentation and without creating any negative impact on the workforce. Thus, the prioritization of business process areas to be taken for improvement exercises becomes an important task for the group.

Whatever the special group (OEG) has to do in terms of a concrete improvement exercise must follow a method that has been tried out and found to yield the desired results. Improvement efforts are expensive in terms of both time and cost and a hit-or-miss approach or an approach that entails redundancies cannot be adopted. Thus, the group initially formed has a crucial role and top executive leadership has a critical responsibility to evaluate the performance of this group and to take appropriate mid-term steps to make it stay on the desired path.

Many acknowledged experts on management theory and practice have offered sets of principles for an organization to follow in order to join the pursuit of excellence. Of course, these reveal some common elements. Pilcher and Studer (2015) state that organizations achieve excellence when leaders (1) create the right goals with the right accountability and the right leadership development, (2) apply the right approach to manage individual performance and (3) select the right processes, programmes and technology to accelerate the continuous improvement process. The only point where this author would disagree concerns the use of the word 'achieve', since our focus on 'achievement' has been based on the fact that 'excellence' is something to seek for but never to be achieved, especially when 'achievement' leads to some complacency and loosening of efforts. Nine principles have been mentioned by Pilcher and Studer as helping an organization in creating and sustaining a culture that supports excellence. These are

1. Commit to excellence
2. Measure the important things
3. Build a culture around service
4. Develop leaders to develop people
5. Focus on employee engagement
6. Be accountable
7. Align behaviours with goals and values
8. Communicate at all levels
9. Recognize and reward success.

2.7 DATA, TECHNOLOGY AND BEYOND

Any organization sets out (explicitly or otherwise) why it should exist, decides what it would offer (by way of products and services) and to whom (customers, clients, beneficiaries, society at large), operates within an environment from which it cannot be insulated, carries out its functions through its workforce including leaders who have their own respective capabilities, interests and intentions, uses some technology (in a broad sense) and achieves a level of performance which can range from 'below or just required to survive' to 'good' and even 'great'. Every element in this description should be comprehended in terms of several contributors and factors as also the extent to which each of the contributors and factors is likely to affect the eventual performance of the organization. And performance may be influenced by the specific features of the organization.

Data are meant to be duly processed, keeping in mind its possible uses currently or prospectively. Organizations do exist where more data are collected than are processed. Further, data processed and aptly analysed should provide insights into what's happening now, what's likely to happen in future and how to improve upon the current situation. In fact, data are badly needed to replace ad-hoc decisions based on gut feel with evidence-based decisions. A decade back when digital transformation was yet to make a successful inroad into business organizations and the roles of data analytics, machine learning, natural language processing and artificial intelligence had not been fully realized, the nine principles mentioned in the earlier section laid emphasis on human behaviour and human decision-making, which will continue to drive organizational efforts towards excellence alongside effective deployment of technology (in a broad sense).

We should remember that effective use of digital technology rests on the availability of, access to and use of adequate data bearing on the processes being analysed to bring about improvements. Data may be available internally in various functional areas, including data originating from customers and prospective customers, regulatory bodies and business partners, which have been responded to or even not so dealt with.

Data (in the broad sense implying 'whatever are given' to study a phenomenon) are generated continuously through the activities of any organization, consciously and purposefully in most cases and unconsciously in a few. These arise through the interaction that the organization voluntarily makes with customers and business partners and also the reports and responses it has to provide to regulatory bodies and custodians of public interest. Relevant data are a must in such interactions and have to be generated and processed by the organization. Beyond these, data about the business environment encompassing competition and collaboration and data constituting business intelligence providing insights into the ways and means being employed by peers to navigate through challenging problems also need to be gathered. Currently, we have a deluge of data. Of course, data downloaded from any source should be checked to establish that those are relevant to the problem on hand, current, reasonably accurate or correct, free from personal or institutional biases and are adequate for the purpose of making some predictions or taking some decisions based on them. Data capture, data checking and cleaning, data augmentation, for example, through imputation of missing values, data visualization, data grouping or clustering and similar other activities preceding the use of machine learning to obtain meaningful insights constitute a big area that has been unfolding very rapidly over recent years. The roles of data, data analytics, data-based decision-making, data-supported redesigning of processes and procedures along with the effective and responsible application of artificial intelligence have to be duly recognized in any serious efforts in the march towards excellence.

Examples of successful use of data and technology in promoting organizational performance on a sustainable basis should get priority over negative situations in a discussion on excellence. However, to remain alert about likely to be missed-out pitfalls, we can also learn from references to cases where more data are collected than are analysed and more results of analysis are available that are actually used to provide insights or useful inputs for decision-making. The author came across organizations that do collect and process data simply to prepare reports for onward transmission to corporate executives and regulators. Little use of data was made to look at current performance of different processes and to plan for improvements thereon. In fact, a lot of data flow in from customers, sometimes without these being sought after. Such data contain useful information, which should be taken into account for decisions regarding quality, price and delivery of products and services, besides information about specific features causing customer dissatisfaction, which calls for a revision of the design. AI-enabled devices or chat boxes have been being used to identify customer choices, respond to customer queries and complaints and even to apprise them of alternatives and options.

Data, technology and values have to be carefully and assiduously unified in a sincere pursuit of excellence. Along with them, what need to be kept in mind are the vision and the mission of the organization, which may have to be revisited as the business environment changes. This orchestration of data, technology, decisions, value system and functions towards achieving a common goal—already identified by the organization— is the primary task in the search for excellence.

2.8 DATA LEADERSHIP

Activities carried out by organizations (institutions) are often initiated, planned, implemented and even evaluated by data (in the broad sense to cover documents and records, figures and images). Once executed, such activities result in data of various kinds. This is true for individuals also. The data accessed and used by one individual or one organization are only partly created by itself and mostly generated by others. Data can be observed (for example, output features noted after a process has been executed), can relate to some 'desired' state (for example, when a process is being planned) or correspond to some expected or anticipated or targeted state (for example, when a process is being specified). With recent advances in capturing data from different sources, in different locations and at unimaginably high speed, organizations find themselves in a data deluge. To sift through the huge mass of accessible data for their relevance, veracity and value (for deriving insights into the phenomena of interest) is no mean task. Fortunately, currently available data analytics methods, coupled with machine learning and machine intelligence, take good care of this onerous task, though effective judgement about which data are relevant and can provide actionable inputs for decisions and actions on different processes of interest has to be made by strategic decision-makers.

In the contemporary world characterized by data dominance, the need for developing and maintaining an organizational culture wedded to data has become almost essential in the context of organizational excellence. The concept of 'data leadership' is not just a buzzword but has come to be concretely defined though not uniquely. Advocates of data-driven leadership like Maranco (2024) even speak of 'data excellence', not to mean excellence of data generated, accessed or used, but the effective, discretionary and successful use of data to improve organizational performance. Data leadership goes beyond 'data-informed leadership' to imply 'data-driven leadership'. The roles and responsibilities of leaders—both at strategic and operational levels—who are 'data-driven' do not sacrifice their inherent and acquired capacity to take decisions and initiate appropriate actions, but do enhance that capacity by 'data' and data-derived insights instead of opinions or judgements.

Considering the existing and potential uses of data, we note that data are used to check compliance with (a) internally planned arrangements for various functions and processes and also with (b) regulatory requirements imposed by government, society or even the customer. Enabling comparison with current best practices within the relevant group (of organizations) allows for a potential use of data in the first case. Data in the first case reflect the baseline situation that may justify the maintenance of status quo or may throw up problem areas and the need for change. Data analytics and generative AI may even suggest solutions to the problems to be addressed by the management. Data in the second case are primarily meant to prepare documents as well as to recognize faultlines and to initiate improvements.

Data-driven leadership calls for a clear and comprehensive data strategy to spell out the purposes for which data are to be collected and the types and sources of data, how and where are the data to be stored to maintain privacy and security (in clouds or data warehouses), how are the data to be processed for getting insights into problems and trying out suggested solutions keeping in mind resources available for the purpose and what will be the end uses of outputs from data-derived 'stories'. A good data story explains what's happening currently (the baseline situation), why does it matter (the need for change) and what can we do about it (the data-suggested change actions).

Data leadership is an integrated approach to leadership, strategic as well as operational, data management and change management. Apart from a data strategy, building the real data expertise to effectively implement the strategy by taking into account the current capabilities and top leadership involvement in developing a data-oriented culture across the organization are essential components of data-driven leadership.

Data leaders are not merely data providers, data processors or even data scientists who work in isolation from those who identify problems, try out solutions and provide feedback on the results. A data leader must be given ownership to problem domains and should be a part of the strategy team. A digital community connected through an effective e-communication network, an adequate IT architecture, involvement of executive management from all functions and inclusion of the use of technology in business and in meetings and discussions are needed to establish data leadership.

Combining the ingredients of data leadership—leadership abilities and traits, expertise in data science and change management—is possible only through a closely knit team of professionals in respective areas with a commitment to work together with support from an inspiring top leadership. When these three ingredients properly work in sync, it can go a long way in boosting operational performance and leading the organization on the path to organizational excellence.

3 Measuring Organizational Performance

3.1 NEED FOR MEASUREMENTS

Management at all levels in an organization is concerned with the 'performance' of the individuals working in and for the organization. The overall performance of organizations is a matter of concern for the related sector of the economy they operate in. The performance of different sectors of the country's economy eventually determines the country's performance on the economic front. In this context, the assessment of performance becomes a crucial task of the management. Assessing or measuring performance, except in simple situations, poses several problems. The challenges in measuring performance are as varied as the organizations themselves—from manufacturing industries in the private sector to publicly funded research organizations to public sector organizations providing various types of services.

The word 'performance' has two distinct meanings: (i) a process of performing, executing or carrying out a task or discharging a responsibility and (ii) the output or outcome or impact of this process. As a process, performance can be assessed right during the process or shortly after the process is completed, while going by the second definition, performance may become relevant after quite some time. A student coming out of an examination hall is asked by an anxious parent or guardian "How did you perform?" and the answer, honestly, would be an assessment by the examinee of how did they answer the questions. Later, during an interview, the same person is asked by an interviewer, "How did you perform in your graduate-level examination?" The answer would be in terms of a score or grade awarded to the examinee by an assessor. The process leads to the outcome, though not always uniquely or predictably. Further, a self-assessment of the process of performance (as in this case) may not be in sync or consistent with an external assessment of the outcome of the process. The first, of course, is subjective. The second also may not be completely objective.

A consideration of performance is important both in respect of single individuals (working within and/or for an organization) and for the organization or of a segment of the organization. An assessment of performance (during a completed period) for an individual member, a group or business vertical (a team, virtually) or even the organization as a whole acts as an important input for making decisions about retention, promotion (in the same or a different task) or retrenchment of an individual member, dissolution or re-constitution of a team, as well as for continuation with or change in

DOI: 10.1201/9781003591733-3

the activity profile of the organization. The most important objective for measurement of performance is to identify the areas of improvement, verifying the effectiveness of the existing strategy in meeting the goals of the organization, fixing accountability of individuals and teams, in benchmarking the organization's performance in a given dimension against industry peers acclaimed for high levels of performance and for knowing the perceptions of concerned people about the organization and its activities.

'Performance'—especially in the context of an organization—may relate to (1) activities carried out or processes performed by an organization or (2) outcomes of such activities or processes in relation to the overall goals to be achieved. Processes include people management, resource management and interactions with the environment. Outcomes could include financial parameters, stakeholder satisfaction, impact on society and impact on the environment, among others. Each of these outcomes can be judged using multiple measures.

With regard to outputs, outcomes and impacts, bringing out distinctions among these terms may be helpful in some contexts. That way, outputs are generally measurable more or less on the completion of the processes, while outcomes and particularly impacts may not be immediately visible. The outcomes and outputs may be sometimes spread over a time span and may even be observed only after a reasonable time frame. Whenever we have to assess such outcomes or impacts, which are not amenable to an immediate measurement, a number of problems arise in developing an appropriate way of measuring them.

Performance—whether in relation to a process or its outcome—can be absolute or relative. Most often, performance is assessed against certain norms, standards, specifications or targets. For certain processes, such specifications can be conveniently and objectively laid down, while for some processes, it may not be as easy to develop and deploy such norms or standards. Targets such as quantity to be produced, time to complete a project or the volume of sales by a certain time and the like can be set as strategic decisions for improvement in process performance and as outcomes or results, they are the bedrock of excellence. It is true that not all aspects of process performance and results achieved can be measured on a numerical scale, though they can also be assessed on an ordinal scale, say ranking. Quantification of manifest variables like revenues, costs, time units, workforce strength or assets created and maintained, energy consumed and a host of similar features or aspects is quite common and each of them can be measured by more than one quantifier. Latent variables like customer satisfaction, impact on the environment, reaction in the society or trust of stakeholders and the like can be quantified only in terms of substitutes, approximations or intelligent conjectures. Research on the quantification of latent variables and the relations among themselves or between them and manifest variables has been growing. Notwithstanding the problem of quantification, measurement to enable quantification has been advocated by social scientists and, in some situations, even by physical scientists. The role of measurements in the context of improvement has been emphasized in the following oft-quoted lines:

If I can define it, I can measure it.
If I can measure it, I can analyse it.
If I can analyse it, I can control it.
If I can control it, I can improve it.

The pronoun 'it' stands for any noun, any phenomenon, feature or aspect of interest to us. In this chain of activities, improvement comes at the end and starts with measurement, which is preceded by the task of concretely and clearly defining the entity to be measured and finally improved.

A proper measurement system yielding measures in respect of each process variable and each variable reflecting an outcome or result achieved is an absolute necessity for working out an improvement in some chosen aspect(s) of a product, a process or a system. Subsequently, the extent of improvement and the level of excellence can be assessed. Barring the simple situation to find out whether or not a certain target could be achieved, such as adopting a new technology, entering the export market for the first time, developing a new delivery mechanism and the like where a binary assessment is made in terms of 'yes' or 'no' response, we need some form of measurements to indicate something close to that target. In a majority of the cases, we need values, scores or ranks to track improvements over the baseline measure.

Such measures must be generally dimension-less or unit-free and should be interpretable in terms of progress towards improvement. These should be data-based and free from personal biases. Since individuals within an organization carry out differentiated tasks, it will be desirable to devise measures of task performance by individuals. Of course, measures of organizational performance are of primary interest to us, but there is no obvious way to link the measures of individual performance to arrive at an acceptable measure or a set of measures reflecting organizational performance.

Needless to assert, computation of any such measure(s) by itself may not mean much, unless it is analysed in relation to other measures, interpreted to imply certain corrective or value-enhancing actions and finally implemented by management in the concerned area(s). A somewhat disappointing situation arises when more measures are assessed than are interpreted in terms of recommended actions and more recommendations are made than are finally acted upon.

Some of these issues have been briefly discussed in the following sections.

3.2 PERFORMANCE OF AN INDIVIDUAL

When different individuals work within an organization, measurement of their performance becomes a tricky task for various reasons. Apart from differences in the intrinsic difficulty or complexity levels of different jobs being pursued by different workers, differences in the environment encompassing all forms of interactions with superiors, colleagues and juniors also affect the performance of different individuals. The same is also true for different groups of individuals entrusted with different types of jobs, placed in different environments and expected to achieve different objectives.

Performance can be measured somewhat inversely in terms of the frequency or magnitude or consequence (maybe in terms of the associated cost) of some indicators of 'poor' performance. An inherent assumption in this approach is that there exists (even though not always communicated) a standard operating procedure to perform or carry out the process and that deviations from the specifications given therein can be identified and quantified to the extent needed. The following is a list of seven generic indicators and only those which are relevant in the given context should be considered.

1. Defect, that is, deviation from the relevant specification given in the process plan regarding any observable feature(s) of some items or elements in the process
2. Rework—when the deviation can be corrected by repeating some work item(s)
3. Reject or scrap—when the deviation cannot be corrected
4. Late delivery—in terms of the work output arriving late for the next work item or process
5. Work behind schedule—may be caused by the delayed completion of the preceding work item or otherwise
6. Lost items—when the output of some work item(s) was not properly preserved or saved and could not be retrieved when required.
7. Items not required—in terms of some work item(s) not relevant in the process or are redundant.

It must be noted that besides the above indicators which are definitely useful in the context of manufacturing and certain types of for-profit service organizations, we should think of other indicators that are appropriate and important in the context of, say, academic institutions or R&D establishments. Thus, in the case of a teaching institution, we may have to consider the processes performed by faculty, administrative governing body and support staff as distinct entities and performance indicators—negative or positive—have to be separately developed for each category. Most of the indicators listed above may apply to support staff members and even to a section of people engaged in the administration. However, these will not apply as such to those engaged in teaching. Joining classes behind schedule, not being able to take questions from students and addressing those adequately, improper assessment of students' progress in studies, incpt handling of rough situations within the organization, failure to provide additional care to students lagging behind others in learning and comprehension, failure to provide right inputs to governing body for improving courses and curricula, augmenting facilities for learning and self-development and the like are some suggested negative indicators of performance for members of the teaching staff. This is an area where much discussion is ongoing.

The development of performance indicators for individuals engaged in different broad categories of tasks in a research and development organization is quite a challenging job. Gaps in mental abilities required to work in a team, identify some mistakes or wrong steps taken during the investigation and initiate proper remedial action, work tenaciously, to avoid any mistake or wrong step detected earlier during the course of the investigation under consideration are just a few suggestions which can be tried out.

In a service organization with a direct interface with users, customers and beneficiaries—current as well as prospective—performance indicators applicable to frontline workers have to include the ability to handle suggestions, comments as well as complaints about services delivered by the organization in a manner marked by courtesy, empathy and respect. Not all these may strictly apply to back-office workers.

A broad distinction has to be made between individuals or groups working in a risky environment (implying that some amount of uncontrollable uncertainty or risk is associated with the outcome or impact of any process performed by such an individual or a group) and others who face little or no uncertainty and can conveniently relate their efforts to the outcomes or impacts they target to achieve.

3.3 ORGANIZATIONAL PERFORMANCE

Before embarking on the task of developing suitable measures for each important dimension of organizational performance that can be retained as a vector of metrics or combined somewhat artificially into one single composite, we have to broadly agree on a definition of organizational performance that is amenable to measurement. As noted earlier, this entity is multi-dimensional and within each identified dimension there exist several features which justify different measures. Generally speaking, organizational performance may be defined to suit the need for such a measurement. This need could be to (1) identify weak points in the current state of affairs relating to people, processes, markets or investments, (2) institute comparisons across functions and teams, locations and markets, (3) address environmental and social concerns raised by the civil society or government or other regulatory bodies or (4) guide the management in achieving the vision and mission of the organization. We may well argue that we need measures that are meant to identify weak links or to guide the management towards improvement in performance. The focus need not be on getting an award for excellence.

Performance can be measured at different levels for any dimension and we want a measure at the aggregate level applicable to all that the organization is required to do, all that it does and all that it achieves during a certain reference period compared to the previous reference period(s).

Interestingly, as many as 207 measures of organizational performance have been noted by Richard et al. (2009) in a review of 213 articles published on the topic during a period of only three years. Added to the bewildering variety are the demands of different types of organizations, from purely commercial enterprises in the private sector to public sector service enterprises and academic or R&D institutions and charitable trusts. Against this backdrop, we note certain areas on which all researchers agree on capturing organizational performance. These are the economic, operational and human capital performance, with the more generic 'economic' performance being split into financial accounting outcomes and capital market outcomes.

To begin this section, we recall an early attempt to assess organizational performance attributed to K. Ishikawa, the father of quality movement in Japan. To evaluate the performance of individuals working in and for an institution as well as the performance of the institution itself, we may take due advantage of the Ishikawa model given in terms of a few equations:

$$\text{Knowledge} \times \text{Skill} = \text{Ability}$$

$$\text{Attitude} \times \text{Environment} = \text{Motivation}$$

$$\text{Ability} \times \text{Motivation} = \text{Individual Performance}$$

$$\text{Individual Performance} \times \text{Institutional Resources} = \text{Institutional Performance}$$

This model explains individual performance as the output of four determinants:

a. Knowledge and skill possessed and applied by the individual for discharging job responsibilities
b. Attitude towards the jobs assigned as challenging versus routine and monotonous, providing opportunities for full use of knowledge and skill versus requiring much less level of knowledge and skill
c. Work environment—knowledge to discharge duties and responsibilities assigned and skills to act, react to suggestions and comments and interact with others should be continuously upgraded, even when duties and responsibilities continue to remain the same. Maintaining a friendly and conducive work environment is a responsibility of the organization, which includes the individuals, and thus individual performance is a determinant linked to organizational performance, if we go by this model.
d. Adequacy or otherwise of resources—physical, technological and human-available to the organization.

Organizational performance, according to this model, depends on individual performance and resources—physical, material, technological—available to individuals for gainful use. This early attempt to comprehend organizational performance is rather philosophical, lacking in concrete details of any of the variables involved like skill or attitude that are amenable to objective assessment.

While the above may sound to be a bit idealistic, the elements of an organization as spelt out here are all relevant—maybe to different extents in different types of organizations—and none of these should be disregarded as redundant or useless. In particular, this portrayal of an organization becomes quite pertinent in a discussion on the role of values and ethics on the performance of an organization.

Before proceeding further, we may look at 'team performance' as some entity that influences organizational performance directly. Teams function under varying names in all organizations except the small ones where such differentiations do not exist. The performance of a team takes into account the effects of synergy and guidance (provided by leadership) to enhance the performance of each individual member. In a high-performing team, leadership explains clearly the purpose behind setting up of the team and the goals to be achieved by it. Members possess a diverse set of skills to complement each other and open and transparent communication is encouraged. The leadership provides direction, motivation and support. Members collaborate with one another to capitalize on collective wisdom and take responsibility for respective tasks, besides continuously trying to improve team performance. The performance of a team is also enhanced by assigning the right member to the right job by matching skills possessed by members and skills required for successful performance of tasks, as well as their interests and individual values. Leaders play a vital role in boosting team performance, in tune with the organization's mission and in line with the corporate organization-wide strategy. This strategy is expected to equip each team with necessary and sufficient resources (including technology and training to use the same). One can try to align some of these factors influencing team performance with the elements incorporated in the Ishikawa model.

Given this wide canvas defining organizations and their activities, measuring organizational performance against a multi-element yardstick can justifiably involve complexity. A few of the issues involved have been taken up for further elaboration in what follows.

Tracing the evolution of performance measures and following up—at least to some extent—measures of individual performance mentioned in the earlier section, we can suggest a few measures of organizational performance outlined below:

- Cycle time, considering the time from receipt of the order to delivery at the customer's end
- Time between the release of advertisement or related promotional material for a new product, a new service or a new mode for placing orders and getting deliveries to availability of the product or the service to customers and potential customers
- Response time to act on receipt of an enquiry or a suggestion of a complaint
- Number of repairs and replacements during the warranty period for a product
- Number of notices or advisories or requests from custodians of public interest including the government, the civil society and other regulatory bodies.

The list is not exhaustive and should be augmented appropriately in each specific context.

3.4 CHOICE OF MEASURES

Organizational performance, particularly speaking of business and industrial organizations, has been traditionally comprehended in terms of financial results arrived at using accounting procedures. Even today, financial indicators are taken as the primary measures of performance, though measures of stakeholder satisfaction (including employee satisfaction and customer satisfaction), impact on the environment and even impact on society, quantified through proxy variables, are being considered important in recent times to indicate sustainable performance. A host of financial results like expected profit after interest and tax, return on investment, credit–liquidity ratio, profit per share or gross value added per unit of production can be directly read from the audited statement of accounts or can be computed therefrom. In her book *The Measurement Nightmare*, Smith (1952) points out important limitations of these accounting measures.

Not too unoften, different performance parameters of any organization reveal some form of inconsistency or conflict in the sense that one parameter may behave in one direction and a second in the reverse direction within a particular policy, strategy or operational tactics. Trade-offs are sometimes adopted and the weights assigned to different parameters vary. We have to keep in mind the fact that some of the measures of performance in current use are positive in the sense that higher their values, better is the organizational performance. At the same time, we also use some negative indicators which should be avoided or reduced in the march towards excellence. We cannot simply take the value or figure for such an indicator with a negative sign and add it up with the value or figure for a positive indicator.

An important illustration is provided by measures of safety performance of a national railway system based on the number of train accidents during a reference period. The number of accidents recorded may give a picture very different from the one presented by using the metric 'accidents per million train kilometres'. This metric is more justified to compare safety performance across countries or within a country over the years. However, just the number of accidents carries more conviction with passengers and also with the general public. Both the parties—service providers and users—should be more worried by a third metric, that is, number of casualties inflicted or even a set of metrics devoted separately to deaths, critical injuries and minor injuries. To the organization, a fourth and related metric should be the total amount of compensation paid to accident victims. Another metric of interest to both the parties would be the related to the total time (or better total train kilometres) lost to train movements. There hardly exists a rational way to combine these different metrics into a single index, nor is such an aggregation desired. In order that these different measures can faithfully and effectively portray the functioning of an organization, these should have high information quality (relevance, reliability, etc.) and transparency, verifiability and comparability to yield what is referred to as an accountable business performance measurement (ABPM) system.

Kitchenham (1996) and Winchell (1996) have identified the following properties or features as the main requirements for process performance indicators. These were subsequently discussed by Holweg (2000). The list is generic and should not be considered as exhaustive. Besides these criteria, one should also include the need for economy in data required to obtain any meaningful measure. Acquisition of credible data on any aspect of process performance and/or result achieved therefrom currently or during a past period may not always be a simple and cost-effective proposition. The following need to be taken note of:

1. Performance is not absolute. It is relative to some expected or desired and targeted value or level, which could be specified by the organization with respect to an industry standard.
2. Performance is multi-dimensional, with financial, economic and social concerns duly considered.
3. Performance measures are not independent of one another and are quite likely to be inter-related.

Kueng (2000) points out that even if a process performance measure satisfies the above criteria, it may not be acceptable to the team that is likely to make use of the measure. Therefore, acceptability of the user for adoption in practice is also an important criterion for the choice of a particular performance measure.

Quantifiability: In order to make valid comparisons across time or units or sections and the like and even among peer organizations, any measure of performance has to be quantitative. If performance indicators are not quantitative by nature, they have to be suitably transformed. For instance, the performance indicator 'customer payment attitude' could be transformed into number of days between 'invoice sent', and 'payment received ' or 'invoice paid'. This way, qualitative measures may be replaced or substituted, though not always uniquely, by related quantified measures.

Sensitivity: Sensitivity expresses how much the performance measure must change before a change in performance can be detected. In fact, a sensitive indicator is able to detect even minor changes in performance. It is well appreciated that improvements in process performance will more often than not be marginal, though continuous. Such marginal improvements will not be detected by a measure that is not sensitive enough. Big changes are obvious and the involvement of a measure is not that critical.

Linearity: Linearity indicates the extent to which process performance changes are congruent with the value of a certain indicator. Or, conversely, a small change in the business process performance should lead to a small change in the value of a corresponding performance indicator, whereas an ample performance rise should also lead to a strong change in the level of the performance indicator. This behaviour should be maintained throughout the plausible range of values for the indicator. Otherwise, it will be difficult to interpret the same difference in value of the indicator over different parts of this range.

It may be pointed out that towards the end of the range, the amount of effort and the extent of change to reflect a given rise in the indicator is much higher than the effort and the change when dealing with the early or the middle part of the range of indicator values. So, non-linearity is quite the expected behaviour of the performance measure once an organization has already reached a reasonably high level of performance.

Reliability: A reliable performance indicator is free of measurement errors. To illustrate, if a certain business process has to be rated through a given performance indicator by different experts, the results should not depend on the subjective evaluation of an individual. Inter-rater consistency is an important requisite.

Efficiency: Since the measurement itself requires human, financial and physical resources, it must be worth the effort from a cost–benefit point of view. The measure has to reflect changes in process performance faithfully with a minimum of effort as reflected in the amount of data needed and the type of data processing involved in developing the performance measure.

Improvement orientation: Performance indicators should emphasize improvement rather than conformity with instructions. Therefore, measuring billing errors, number of safety violations, data entry errors and the like do not create an atmosphere where feedback sessions are viewed in a positive, constructive light. Indicators should be so defined and scaled that its values speak of aspects of performance that are directly and not inversely linked with improvement in process performance.

Besides the abovementioned characteristics, the performance indicator chosen should not only reveal the current situation adequately and faithfully, but should also reflect the past trend and indicate the future potential in any important aspect of process performance and results generated from it. Looking at figures for the chosen measures over some past periods, one should be able to predict with reasonable credibility what is going to happen to the organization with respect to its performance.

3.5 DIFFERENT APPROACHES TO PERFORMANCE MEASUREMENT

There has been a growing tendency to use generic approaches, measures and tools that are thought to enjoy universal applicability, without focussing on specific features of individual entities or types of similar entities. This tendency, arising out of the globalization phenomenon, has its merits and limitations. It highlights even conceited or artificial similarities at the cost of genuine and important specificities.

Comparing the measures of organizational performance and yardsticks against the goals and objectives in a charity trust cannot be the same as those for an educational institution, a business house, or a hospitality organization. In fact, the more important process-specific performance parameters and indicators derived from these processes vary across organization types. Therefore, organization-specific measures or measures applicable to organizations sharing the same broad type of activities and dealing with similar customers or service recipients are more meaningful.

However, key performance parameters and indices based on these parameters apply equally well to a large class of organizations that can be branded as 'business organizations'.

Approaches usually adopted to assess organizational performance consider several dimensions of performance such as the following: (i) achievement of objectives related to growth, innovation, profits and expenditure, relative to goals, (ii) business process efficiency in terms of speed, cost efficiency and perceived agility for both core and support processes, (iii) returns on business investments made and projects undertaken in terms of financial results, (iv) projected versus actual costs measured against the derived benefits, assets and facilities acquired and used.

Marketplace performance is reflected in market share, market size for any new products developed and net sales revenue. Alignment is revealed in the relationship between departments, employees and their departments, organization and its marketplace as well as the organization and the external environment. Workforce involvement can be gauged from their contribution through skills and proficiency, productivity, motivation and engagement, retention and loyalty, support for change initiatives and continuous improvement efforts. Organizational resilience, quite a challenging construct, has also been considered as an important contributor to sustained performance. Sustained performance is a defining characteristic of organizational excellence, as indicated earlier. Approaches taken to come up with meaningful measures of performance both for internal consumption as well as for preparing reports to respond adequately and effectively to mandatory requirements and to inform the society at large about the organization and its activities include use of the following:

1. Key performance indicators (KPIs) for financial as well as non-financial aspects of current performance mostly by way of results pertaining to financial results and customer satisfaction, along with people engagement in some cases.

2. Balanced score card, which is not an indicator or even a set of indicators, but a management tool encompassing how different aspects of the organization have been functioning and what results have been achieved. It is provided in form of a report based on performance indicators, reflecting the strengths and deficiencies of different functional areas that the top leadership may review to take initiatives for improvement wherever necessary.

3. Benchmarking is a comparison of performance in different areas, both in terms of process performance and results generated, against those in well-performing organizations in the same or similar business. Benchmarking is often done against the industry or business leader as the reference.

These apart, financial analysis and analysis of findings from special surveys and feedback and social responsibility reporting are sometimes taken into account for an understanding of organizational performance.

3.5.1 Use of KPIs

A KPI may be based on a single parameter characterizing a single operation, process or business segment, or it could be derived from several parameters relating to a single entity or even to several inter-related business processes. KPIs measure performance as is currently realized during a specified 'reference period' against the corresponding goal or target set by the organization. These indices are meant to reveal the observed trends in performance in the areas under consideration and to provide an indication of prospects as well as problems in the future. In some sense, these even help the leadership to revisit the goals and targets as and when necessary to ensure that such goals and targets should be achievable, while remaining ambitious. This three-fold deployment of KPIs has been being adopted by more forward-looking organizations since the close of the last century. Among the more widely used KPIs the following five are commonly found:

- Revenue growth (gross and net, over the last reference period)
- Revenue per client (not so important for the fast-moving consumer goods (FMCG) sector). Clients could mean distributors or agents or other intermediaries)
- Profit margin (can be computed in several different ways)
- Client retention rate (should also take into account induction of new clients separately)
- Customer satisfaction (involving customer perception about quality, price and value)

In the category of KPIs for financial performance, several rates and ratios are being used in some organizations suited to their specific needs. For example, the expense ratio is obtained by dividing a fund's (a mutual fund or an exchange-traded fund) operating expenses by its net assets.

The main purpose behind use of KPIs is to help management take informed decisions regarding disaggregated business processes for which different individuals, groups or teams are accountable so that the corresponding goals or targets can be met. The use of KPIs increases accountability and avoids 'passing the buck' to someone else for a shortfall in performance. The measures used by an organization to assess its performance may not always comply with those of the 'environment', as they relate to 'what is right for the society' and not simply 'what is good for the organization', while interacting with or overseeing the performance of the organization.

There have been many constructs about the holistic performance indices (HPIs), which capture both financial and non-financial performance indicators or success factors. Besides measures of overall performance, we should also involve measures of the performance in sectors that pertain to the employees, customers, business partners, including suppliers, and society.

Sectoral performance indicators include the Customer Satisfaction Index, Employee Satisfaction Index, Expected Performance Rating Index, Reputation Excellence Index and Society Trust Index. These indices take into account certain 'manifest' variables and then proceed with some statistical analysis using a structural equations model to derive measures for the outcome variables that are latent.

Traditionally, performance has been reported in terms of financial metrics based on costs incurred and revenues earned. The focus was on markets and customers, on the one hand, and investors, on the other. Quite a few ratios like value addition per worker were being calculated to track the performance of an organization and these would reflect generally on the last completed reference period (mostly a year and sometimes a quarter) either as an absolute figure or as a percentage of the corresponding figure for some previous reference period(s). Detailed analyses like analysis of cost of poor quality, cost of maintenance or value addition per worker were also being worked out. Explanations for the observed measures of performance in terms of controllable features of the organization and its activities were not paid much attention. Therefore, the measures of financial performance by themselves were not adequate for the purpose of looking ahead and planning for the future.

3.5.2 Balanced Score Card

It was being gradually realized that summary figures like indices were inadequate even for the purpose of appraising the current level of performance—overall or for a specific functional area. Such figures would conceal more than what they revealed. Aggregation would often lead to inadequate attention paid to an area where performance parameters taken individually would reveal an unsatisfactory state of affairs, but put together with values of parameters in certain other dimensions, they would present an otherwise satisfactory picture. Thus, for a wholesome review of organizational performance, there was a dire necessity of putting together objectives and targets against which performance measures have to be assessed, along with measures of current performance and even steps initiated to improve matters which are under way—all in a single report to facilitate the top leadership to relate performance in one dimension with those in another and to initiate measures to align performance with the organizational vision and mission. Such a report based on adequate quality data collected and duly processed can enable top management to revisit the mission, examine deviations from goals and targets set therein and and to revise targets wherever needed.

Before we proceed to discuss briefly the widely known balanced score card in the next chapter, it will be useful to remember that the measures or scores for various dimensions and sub-dimensions of organizational performance, especially in respect of non-financial measures, are essentially derived from some survey of perceptions, beliefs and suggestions linked to some manifest variables and converted to a score. The validity and utility of the score depend extensively on the quality of responses obtained from the survey, normally marked by fallibilities like non-response, response bias, response wrongly elicited by the investigator, incorrect responses which are unintentional and stem from ignorance or lack of information and a host of other contributory factors. It has to be admitted that any deficiency in the data cannot be compensated by any amount of sophistication in data analysis that is used to provide an index of performance.

3.5.3 Benchmarking for Excellence

Benchmarking has been being used as a general business practice since around 1980, starting with IBM and Xerox in the United States. The urge to perform at least at the same level as the competitors was reflected in the form of benchmarking. Benchmarking has benefitted quite a few manufacturing and service organizations in improving their business processes and performance.

The development and realization of improvement objectives in business processes and their outcomes, particularly 'stretch' objectives, is often aided by the mechanism of benchmarking. Usually, process benchmarking and performance or results benchmarking are considered as the two important areas where this practice has been adopted. Some people talk of benchmarking competitive advantages and strategies.

Four philosophical steps in process benchmarking have been mentioned by several authors: (1) know your operations; (2) know the industry leader or competitors; (3) incorporate the best practices and (4) gain superiority. Camp (1995) refers to four types of benchmarking: internal, competitive, functional and generic. Internal benchmarking focuses on best practices within an organization across different areas, divisions or departments. Of course, benchmarking for improving process quality mandates the same or similar processes to be carried out in the best-practices scenario. Competitive benchmarking provides a comparison among direct competitors, most of whom may not like to share information regarding distinctive key features of their processes. Functional benchmarking is done across units within the particular industry or even outside the industry. Generic benchmarking focuses on innovative work processes in general.

The formal definition for benchmarking could be any of the following:

- The continuous process of measuring products, services and practices against the toughest competitors or those companies recognized as industry leaders
- The search for and implementation of best practices
- A systematic approach by which organizations can compare themselves in certain selected areas or dimensions of their operations or results against the best-in-class organizations, can measure and analyse gaps and initiate necessary improvement actions

Various types of business information can be gathered, derived from primary data collected through direct interactions with benchmarking partners as well as from secondary data gathered from published documents or put in the public domain. New business knowledge can be created by analysing specifics of various business factors of competing companies, especially those known to perform better, and comparing those with corresponding factors in the present organization.

This knowledge can be used to come up with new decisions for controlling the relevant factors and develop actions to improve processes and results.

The goal of process benchmarking is to gain knowledge about the characteristics of planning, designing, executing and controlling and even evaluating various business processes and activities involved in the processes selected for improvement by which competitors successfully implement their strategies. The goal is to improve the process(es) in the given organization:

- Since benchmarking involves comparison with other competing companies, an important problem is to choose the right companies against which to benchmark the selected process(es). The general advice is to consider
- Other companies within the same group—internal benchmarking
- Competitors within the industry
- Other companies in the industry, which are not direct competitors and other companies in other industries.

The last one is more applicable in the case of process benchmarking where a particular process like space utilization or interacting with customers in an industry that offers a completely different product or service profile but is known to perform excellently well in the selected process. Of course, such a choice of the benchmarking partner requires a deeper knowledge by decision-makers in the organization about the given process and also the organizations which are required to perform this process and which have performed with a high level of efficiency.

3.5.3.1 Some Early Cases

In the 1980s, Xerox studied their direct competitors among Japanese companies to discover the following:

- Unit manufacturing cost equalled the Japanese selling price in United States.
- Number of vendors was nine times that of the best companies.
- Assembly line rejects were ten times higher product lead times were twice as long defects per hundred machines were seven times higher.

Xerox started process benchmarking seriously and effected the following changes:

- Number of vendors was reduced from 5,000 to 300.
- Concurrent engineering practice was introduced.
- Commonality of parts was increased from 20 to 60—70 per cent.
- Cross-functional teams were put in place.

The results were surprisingly huge: quality problems were cut by two-thirds, manufacturing cost cut by 50 per cent, development time reduced by two-thirds, direct labour cost dropped by 50 per cent and corporate staff reduced by 30 per cent.

One of the earliest benchmarking exercises recorded in manufacturing industries was the one taken up by Motorola in 1998 and subsequently by several industry majors in the United States and in other countries. Different models for benchmarking have evolved over time. The basic functions in benchmarking are (a) choose the business areas to be benchmarked—some core or support process or results and (b) choose the benchmarking partner. For obtaining optimum results through benchmarking, the partner is usually an organization engaged in similar processes and products and comparable in size. However, for process benchmarking, the partner may belong to a different class of organizations, carrying out a process with recognized effectiveness and efficiency.

3.5.3.2 Elements of Benchmarking

It is quite a difficult task to choose different benchmarking partners to improve efficiency, effectiveness of several business processes and to initiate actions to bridge the gaps in performance. If breakthrough changes are suggested to cover big gaps, top management's support to implement such changes will be essentially needed.

In essence, a benchmarking exercise provides inputs for process improvement, and in most cases for breakthrough improvement. These inputs along with others like findings from failure mode and effects analysis (FMEA) have to be converted into feasible corrective and preventive actions to improve process quality. Actions being used in the best-in-class organization may not apply mutatis mutandis to your organization and the task is to work out suitable modifications so that such actions are suitable for your organization. The evaluation of results on the completion of one round of benchmarking is a must to ensure that the organization remains where it was at that point in time.

Creative elements have to be injected within and beyond the current practices prevailing in the concerned industry to achieve improvement in a situation where most of the units or organizations have already reached their limits of technological and managerial competence. These extensions of benchmarking require serious creative thinking or breakthrough followed by appropriate actions. This implies the need for thinking out of the box, as opposed to thinking normatively (finding an existing best practice and adopting it with or without modification).

Brainwriting and brain-mapping exercises have considerably broadened the scope of brainstorming and its applications. It may not be wrong to mention here that advocates of early pragmatic and conventional ways of forming collective views have called brainstorming simply an 'old wine put in a new bottle'. A related and modified exercise known by the acronym SCAMPER seems interesting and actionable in the context of benchmarking. This mnemonic, in the expanded form, refers to several elements:

- Substitute (the existing or the present entity by another)
- Combine (several entities currently considered into a broader entity)
- Adapt (a new entity to the present needs and situations)
- Modify (an existing entity in respect of its relevant features or/and functions)
- Put to other uses (instead of the current less-value-adding use)
- Eliminate (an existing entity or some of its features or functions that are irrelevant or undesired)
- Erase (dilute some features and functions)
- Reverse or re-arrange (components within the existing configuration).

An entity, in this context, carries many meanings. It could imply a hard material object or a component within a product, or an activity/operation within a comprehensive process, a procedure, a document, a design or a test/experiment. Accordingly, the various elements of SCAMPER have to be interpreted and implemented.

3.6 PERFORMANCE MEASURE IN THEORY OF CONSTRAINTS

The theory of constraints (ToC) considers an organization as a set of interdependent processes that are required to work in harmony to achieve the profit goal for the organization as a whole (Gupta, 2002). It therefore looks for total system performance rather than localized measures for guiding the management. In fact, an organization's success will depend upon the compatibility between the performance measurement system and the organization's global goals (Lochamy III, 2002). Among the results achieved by an organization to be judged against any goal or target, ToC considers only financial results and a few results that are not always taken into account in the usual measurement of performance on the financial side.

Goldratt and Fox (1986) prescribed two sets of performance measurements: global measurements and operational measurements. These measurements are based on throughput accounting (TA) paradigm, which significantly departs from traditional accounting practices. It provides an alternative way of analysing the economic health of the organization that eliminates harmful distortions that might stem from the existing accounting practices. It also helps managers in understanding the contribution of constraint or constrained resources to overall profitability. Thus, managers with the help of TA practices can identify and leverage constraints to maximize the profit for the organization.

The traditional accounting viewpoint does not support the goal of improving throughput in the long run. The traditional approach considers inventory as an asset that leads to the inflation of an asset and generates a false sense of making a higher profit. Accumulated inventory may not be sold sometime either due to obsolescence or market fluctuation. However, in this way, material inventory gets disguised as a part of operating expenses. Consequently, the organization would not earn real profit despite holding a large inventory in the store. Therefore, it may lead to paper profit and promote undesirable behaviour. Secondly, it emphasizes maximizing the throughput rather than cutting expenses. The strategy of expenditure cut stops working when expenses reach zero while increasing throughput has no limits. TA uses three core measures—throughput, inventory and operating expense—which can be controlled and manipulated to increase the contribution of constrained resources to overall profitability. To increase the throughput in any organization, the following measures can be taken: (a) increase throughput, (b) decrease inventory and (c) decrease operating expenses. The concept of throughput, inventory and operating expenses are explained in the following paragraphs.

Operating expense: Operating expense (OE) represents all the money that is spent to convert the input into throughput or to generate goal units. In the throughput-world paradigm, it is the cost of maintaining a facility at a given level of capacity. This does not include expenses like cost of raw material, commissions and electricity charges that vary directly with manufacturing or service volume. However, it covers expenses incurred on items like rent, salaries and insurance fee, which will be paid even if no operation is carried out for a while. As such, operating expenses can be defined as all the non-variable costs associated with turning the inventory into throughput (Mabin and Balderstone, 2003). In simple terms, it can be expressed as follows:

OE = The total cost of maintaining the system (i.e. rent, lease, labour, vehicle, materials and equipment, insurance fee, etc.)

Throughput: Throughput is defined as the rate at which goal units are generated from customer sales, not through production. It is mathematically expressed as a rate of goal unit per given time period like day, week or month. If the goal represents money, throughput will be an amount of money generated per unit time. Accordingly, it is computed as revenues received minus total variable costs for the chosen time period (Blackstone 2010b). It therefore becomes imperative for marketing and sales departments to be aware of the throughput dollars per constraint and push these products in the market and not the ones having the highest selling price or margins (Blackstone, 2001). Throughput can be considered as

T = (Total Revenue – Total variable expenses)/Time period

In reference to throughput world thinking, productivity (unit cost) can therefore be calculated as:

Productivity = T/OE

Inventory: Inventory is the money locked in goods that an organization intends to sell or the material which it intends to convert into saleable goods. In the case of operation management, inventory refers to those stocks or items which are used to support production (raw materials and work-in-process), supporting activities like maintenance, repair and operating supplies, and customer service like finished goods and spare parts (Blackstone, 2010). If efficiency as a performance measure is focussed on non-constraints, it might lead to many dysfunctional behaviours, that is, accumulation of unneeded inventories in many forms at multiple points. Material requirement assessment, procurement, allocation and ensuring its availability in market form one of the most important project tasks. Therefore, material management in the project execution, which include material testing, procurement, inventory control and availability, shall be given due importance. In essence, inventory is considered as all the financial resources invested in purchasing things the system intends to sell (Lockamy III, 2003).

ToC uses another set of global performance measures: net profit, return on investment and cash flow. Net profit is an absolute measurement which is obtained by deducting OE from throughput and mathematically expressed as

Net Profit = T – OE

Return on investment (ROI) in throughput accounting paradigm. ROI is mathematically expressed as

ROI = Net Profit/ Investment

where Net Profit = Throughput – OE.

Investment represents money that is tied up in all physical things like work in-process inventory, finished good inventory machinery, real estate and equipment. In earlier ToC literature, investment and inventory were used interchangeably.

Cash Flow: Cash flow is a red line of survival, an 'on–off' type measurement (Rahman, 1998). This measurement becomes important when an organization faces the issue of cash crunch, while it loses importance when there is no cash crisis.

3.7 PERFORMANCE OF ACADEMIC OR R&D ORGANIZATIONS

As can be well imagined, the performance of an academic institution, where customers are potential beneficiaries, including students and their guardians and potential employers and even the society at large, will be quite different from measures that apply to manufacturing or trading organizations. Similarly, customers, beneficiaries or potential beneficiaries in respect of a research organization or an R&D establishment may be difficult to identify clearly and unambiguously. A private R&D establishment may take up some sponsored research (mostly applied) or the development of a simple prototype, though processes to be performed during the course of the work involves uncertainty and repeat trials. For a publicly funded R&D organization, stakeholders range from funding bodies to collaborators and partners to society and economy at large. Added to the problem of identifying the customer/user profile is the fact that the expectations and needs of the customer or beneficiary groups are unlikely to be in the form of concrete statements. In fact, students in an academic institution are most often not even allowed to communicate their choices about courses and subjects as well as about take-aways.

It is sometimes argued—and quite rightly—that an institution of higher education like an undergraduate college, not to speak of a university, must promote knowledge creation by its faculty and researchers beyond simply teaching or communicating knowledge. In fact, teaching and research in such cases are inalienably linked to each other. To perform research activities in educational institutions and in organizations in the public or private sector, which carry out sponsored or open research—in some priority or thrust areas identified by a national body—requisite facilities and proper atmosphere are essential for research pursuits to be effective. In the following two sub-sections, we will focus on measuring or rather assessing research results.

3.7.1 Case of Academic Institutions

For an institution in either the public or the private sector, quality of the teaching–learning process, adherence to accepted and/or mandated norms for the admission of students and recruitment of faculty, maintenance and regular augmentation of library and laboratory facilities, besides provision and upkeep of basic physical infrastructure, including necessary conveniences, illustrate process performance. On the 'results' side, the performance of students in internal and external (public) examinations, participation of faculty and even the non-teaching support staff in internal development programmes and in external events like conferences, seminars, debates, discussions and lectures and percentage of students joining a placement camp who are eventually selected by prospective employers are examples of results targeted and achieved.

Dealing with the quality of teaching or the performance of a teacher, Mukherjee (1995) points out some limitations of the Ishikawa model and claims that a band of dedicated teachers makes up for scant resources, that the profound knowledge of a teacher creates the desired effect on the students despite the former's unimpressive delivery of lessons or that keeping its own house in order in the midst of on environment not-so-congenial to the pursuit of knowledge may not be an impossible task for a college or university. The performance of an individual teacher, researcher or administrator on an assigned job does not depend merely on their ability to perform the job but also on the extent to which they feel motivated to do the job. In this context, Mukherjee reminds us of the fact that programmes to help individuals in refreshing and augmenting their subject knowledge or enhancing their skills may also lead to a change in one's attitude towards a given job and thus affect their motivation. A job that seems to be challenging or which matches one's current levels of knowledge and skills may subsequently appear as routine, monotonous or demanding a lower level of knowledge and skill once the individual acquires a higher ability.

The quality of teaching may well be taken as 'quality of conformance'. In the wider perspective of quality, 'quality of design' can be comprehended in terms of policies, plans and procedures to be followed by the organization in regard to admission of students and their promotion/retention, recruitment and engagement of faculty and administrative and support staff, besides teaching aids and accessories and facilities for study and work available to students and teachers. Even the choice of courses offered and teaching and evaluation methodologies to be adopted are important elements of this 'quality of design'.

The quality of performance of an academic institution is comparable to outcomes or results in the context of organizational performance, while quality of design and quality of conformance are analogues of drivers or determinants of performance. Some institutional ranking frameworks exist at national and international levels with the primary objective of assessing performance, with a focus on results or outcomes. The outcomes are stretched beyond the performance of students in public examinations or on the performance of graduating students in their respective work areas to indirectly in the responsibility and remuneration profiles. The frameworks usually include perceptions of employers as well as of senior academics about the reputation of each of the institutions participating in the ranking process.

3.7.2 Case of an R&D Organization

Research performance is a debatable concept. It may be taken as 'quality/value' of research, without reference to any pre-specified standards, or as 'excellence' in terms of conformity to some accepted superior standards. It is multi-dimensional, related to the research process and to the research findings/output, if not to the research objective(s) also. Each of these processes and outputs is to be judged (given the objective(s) or expected outcome(s)) in terms of several relevant aspects. Different criteria reflecting different points of view or concerns may have to be involved.

Performance in the process takes care of the appropriateness, adequacy and novelty of the research design in relation to the stated objective(s) and of conformance to design requirements during each step in the research process.

Performance related to results/output is based on the way results have been validated and on the nature and extent of knowledge created through the results or in terms of the value or worth of 'concrete entities' (systems, processes or products) generated through the research.

Research performance as an overall concept cannot be 'generic' and should be 'specific' to the domain of research and the research objective(s). It may, however, involve a few elements common to a wide class of research.

One can think of the domain in which research has been done or the context of the motive(s) and objective(s) as defining some category within which the 'performance' of research can be assessed. A research with limited objectives or expected outcomes cannot be placed in the same bracket as prolonged research with ambitious and comprehensive objectives. It can be appraised or assessed, maybe on a 'semantic differential scale'. The Likert scale is also a widely accepted scale. Appraisal involves experts in the domain of interest. However, it does not usually involve an unequivocal measurement on a numerical scale.

It is 'dynamic' in the sense that appraisal of any dimension, especially the output, has to consider current developments in the domain. Thus the same research work may qualify for one assessment just after its completion and a different one at a later point of time.

The peer review mechanism and the wider approach that takes due account of social, political and ethical considerations of the research process and the worth or value of its output provide two complementary (and not alternative) approaches to judging the value or excellence of a given research.

The first approach, with a proper identification of 'peers' and an imaginative identification of merits and deficiencies pointed out in their reviews (within a specified period from communication of the findings), takes care of both the process and the output.

While the first approach is more 'global' in character, the second may justifiably reflect 'local' concerns and requirements. In fact, both the approaches should be used since peer reviews of outputs from engineering and related research may not necessarily take account of 'local' needs as well as the potential impacts of the outputs.

In either approach, we can make use of some numbers or grades for each aspect—of the process and the output—being appraised. These could be standardized to yield some 'indicators' and subsequently the different indicators can be combined using suitably developed 'weights' to yield a composite assessment of performance. Development of indicators and their weights usually poses problems.

Speaking of the first approach, we often use counts (of publications, citations, patents, algorithms, awards) as are applicable in a given situation, assign some weights subjectively to each count and take the weighted total to make an assessment of performance. Not much is done to discriminate between discernible categories within the underlying attribute yielding a count.

A publication dealing with a new topic or new problem or, say, using a new methodology should be rated higher than the one that adopts a known methodology and treats an existing problem, maybe from another perspective.

Publications by an individual or a team may, on one extreme, address the same type of problems with more or less the same methodology to add more knowledge about the same phenomenon. The other extreme may correspond to a case where each publication deals with a new problem, maybe within the same broad domain, using the same or different models, methods and techniques. Between the two extremes lies the big majority of publications. And it is quite a problem to judge which extreme is better.

Citations may correspond to a wide variety of realities and can be placed in different categories, such as the following:

1. Just mentioning the cited publication as some previous work in the domain
2. Pointing out some limitations or infirmities in the cited publication
3. Extending the scope and coverage of the work in the cited publication
4. Deriving motivation to address a new problem
5. Drawing conceptual, methodological or technical support from the cited publication to create new knowledge.

Evidently, these categories should not be treated on the same scale. Category 4 should attract a higher weight, followed by category 3.

A count of patents granted or algorithms developed and executed through software would be a better indicator. However, their applicability is somewhat limited and the numbers are available only after some further efforts.

A count of awards is not necessarily a dependable indicator, as such recognitions are offered by a number of institutions. Of course, some recognitions by well-established and well-known institutions engaged in promoting excellence in research tend to reflect the true value of research.

Looking from the 'output' point of view, one can think of three discernible aspects such as output, outcome and impact. All these, of course, may not apply in all the situations.

Research output refers to the 'findings' or 'results' as communicated by the researcher or the research team, after being validated in an appropriate manner. In some cases, these could be some algorithms or software, which can find direct application or execution, or some products or processes for production or control, which can be tested for their utility and effectiveness.

It may be difficult to judge the contribution to extant knowledge in the case of theoretical research resulting in some formulae, inferences or methods. Of course, findings from an exploratory research based on a proper research design or confirmatory research supporting or refuting known results can be conveniently judged as contributions to the existing knowledge base.

The second indicator of output is the 'outcome' of the research in either (a) providing a timely solution to a vexatious problem, (b) resulting in a new or improved process, product or service or (c) inspiring and guiding significant research in some unexplored domains. The last one may not be a true reflection of the results or findings. An 'outcome' distinct from results or findings may not emerge from all types of research.

A very important indicator of research performance that may require elucidation in a manner specific to the research work under consideration is in terms of its 'impact' on the search for new knowledge as well as on society, economy and environment.

A change in social outlook, the consumption of non-renewable natural resources, policy-framing exercises or the dissemination of vital information among members of the public that result from the adoption of results and recommendations of research illustrates its real 'impact'.

Given that 'impact' can be assessed with a fair bit of credibility for all types of research, it will be worthwhile to concretize the concept of 'impact' and develop a scoring system so that scores can be assigned to each identified and relevant aspect of 'impact' and weights for the different aspects of 'impact' assigned so as to arrive at an overall score to assess research performance.

Different types of research justify and demand the use of different criteria for assessing their worth or value. The value of any given research should not be assessed from only one point of view. An exploratory research can be viewed from the process angle to find the extent to which the findings achieve the expected outcome in terms of identifying different variables—manifest and latent—describing a phenomenon comprehensively and establishing pertinent relations among such variables; it can be also examined from the angle of prospective researches towards controlling and improving the underlying phenomenon that will be motivated by findings of the given research.

A research in the field of industrial management may come up with new ways to enhance the sustainability of profit. It may also provide inputs for better conservation of natural resources. Or, it may open up a pathway for better employee–employer relationship.

Thus, the evaluation of the worth of a research project may often involve different criteria relating to the research process and the research results. The use of a single criterion, mostly in terms of economic benefits, may not suffice.

The entire exercise of appraisal, measurement or assessment of research performance involves primarily three stakeholder groups: developers (of performance indicators), users (policy analysts) and practitioners (including administrators and scholars). Priorities and weights for different criteria to be used in appraisal may vary across stakeholder groups.

The assignment of a rank or a score to a given research work considering the worth of the research process, on the one hand, and outcomes and impacts of research, on the other, may take due advantage of the methods and techniques of multi-criteria decision-making (MCDM). The analytic hierarchy process and several other MCDM methods can be gainfully used.

The performance of a given research exercise on a given criterion should be based on relevant aspects of the exercise and a score or a scale value (on a semantic scale) or by combining several scores or scale values—one for each aspect–if necessary.

Weights for different criteria may be developed through pair-wise comparisons (as in AHP) of criteria or using an entropy measure for the performance scores of the candidate research exercises on each criterion as in the technique for order preference by similarity to ideal solution (TOPSIS).

The involvement of more than one expert or judge presents problems in combining scores or scale values and in aggregating ranks to reach a consensus. It is, of course, possible to check the agreement among multiple rankings of candidate research projects/exercises and to take the average ranks to define a consensus ranking, in case of a sufficient agreement. The problem of developing a consensus ranking per se is an NP-hard problem, where NP stands for nondeterministic polynomial time.

We must remember the old advice, 'one size does not fit all'. The same approach or the same set of criteria may not be appropriate in all cases. Indicators by way of scores or ranks are all fallible, but they are good entry points for discussions or debates on research performance. Such debates should continue in the hope of eventually converging to some consensus measure of performance.

Organizations primarily engaged in development work based on in-house research or on research—possibly on an abstract level or as a theoretical framework at the best tested for feasibility on a laboratory scale—should justify some specific measures of performance. Thus, for research resulting in a procedure, a process or a product, indicators of performance could be the number of patents filed, the number of patents granted, the number of patented outputs productized in-house or in some other concerned organizations with or without any modifications.

3.8 RISK-ADJUSTED PERFORMANCE MEASUREMENT

Some individuals accept risks in achieving their targeted performance, some are exposed to risks while discharging responsibilities assigned to them and attempting to achieve respective performance targets set for them, while some others may be working in risk-free 'certainty' situation wherein a specified level of input results certainly in a specified outcome or output. The assessment of performance by these and possibly other types of workers just in terms of the outcome, output or impact of their inputs should take due account of the level of risk (including zero) involved in each case. Further, such an assessment may also be required to distinguish between situations where individuals are encouraged to take risks and where such risks are incidental to the performance of a task or discharge of a responsibility. In fact, individuals operating at a higher level and required to take strategic decisions are quite often required to navigate through risky situations. The consequences of risk at this level are greater at this level.

As is well known, 'risk' has been comprehended and analysed in terms of its two components: (i) effects on performance (may be quantified by way of a multiplier on the output, outcome or impact identified with performance) of the 'states of nature' representing the business environment (determined by economic, social, political and environmental conditions at micro-, macro- or even meta-levels) prevailing during the performance or execution of the task or responsibility and (ii) the associated probabilities of the realization of such environments as the other. We usually take risk as the product of the (adverse) effect caused by a given environment and the probability that this environment is actually faced during performance.

Not to go any further into the vastly engaging field of risk analysis, it may be simply pointed out that none of the two elements of risk involves a convenient and generally accepted quantification and relevant data to develop a measure of risk are sparse and even not so credible in certain situations. The multipliers are definitely specific to the task and the likely environments, while subjective probabilities associated with these multipliers can be quite misleading.

Assessing the risk-adjusted performance of an individual would mean treading on a slippery path and we may better restrict ourselves to organizational performance. Kaplan, among others, has discussed risk-adjusted portfolio management on the lines of Markowitz using the mean variance approach based on utility considerations. This does not truly reflect risk-adjusted performance of an organization that goes beyond and sometimes avoids a direct concern of portfolio management. We propose a relatively simple approach, which admittedly has inherent problems associated with scenario analysis and probability assignment.

Let S_1, S_2, S_3, ...S_n be the possible states of nature likely to be encountered during performance (as a process) as identified through a brainstorming exercise and let E_1, E_2, ...E_n be the corresponding effects on the outcome or output of performance, taken as multipliers (which could be less than or more than or equal to unity). Let us assume the probabilities P_1, P_2, ...P_n to be assigned to the respective states of nature, based on past data, future predictions and some reasonable relations connecting the state with the outcome. Thus the expected outcome under risk can be taken as $\Sigma E_i P_i = E$, then the risk-adjusted performance of the organization can be reasonably accepted as P / E where P represents a satisfactory combined measure of the organization's performance.

3.9 SOME COMMON MEASURES

Among the more commonly used non-financial measures of performance, Customer Satisfaction Index leading to customer loyalty and customer retention has been discussed comprehensively by many authors. Customer satisfaction is both a driver of excellence and results. Another measure that has drawn attention more recently as a determinant of excellence is the Employee Satisfaction Index, which can be extended to Employee Engagement Index. Other measures of interest include Society Trust Index, Reputation Index, Vendor Rating Index, and the like. Let us briefly discuss the Customer Satisfaction Index and provide only a brief description of other measures. We will, of course, refer to measures of financial performance, which are more objective and data-based in nature.

3.9.1 Customer Satisfaction Index

The measurement of customer satisfaction has been attempted in various ways by different groups, primarily based on responses in a properly designed sample survey of customers or potential customers seeking their opinions of ratings of features and functions of products and related services provided by a supplier. It must be noted that the absence of customer dissatisfaction does not necessarily imply customer satisfaction. There are quite a few variations in sampling, questionnaire design and its administration, scaling of responses, assignment of weights to responses on different items or converting scaled responses or ranks to scores.

The first national customer satisfaction index was the Swedish Customer Satisfaction Barometer (SCSB) designed by Fornell and his associates in the University of Michigan Business School in 1989. The methodology was refined during the period 1989 to 1993 and was pre-tested in the United States in 1993.

Of late, some national standards are used to derive the index from a comprehensive model that includes drivers and outcomes of customer satisfaction. A widely accepted model is the American Customer Satisfaction Index (ACSI) Model introduced in 1994. Based on the SCSB, the ACSI is both a trend measure as well as a benchmark for companies, industries and some sectors of the household consumer economy.

A CSI model, as illustrated in Figure 3.1 consists of a set of latent variables (LVs) that depict the cause-and-effect relation between the antecedents and consequences of customer satisfaction. To measure a CSI model, we have to use a structure equation model (SEM) technique. There are two types of SEM techniques: covariance-based (e.g. LISREL, EQS or AMOS) and component-based (e.g. partial least square).

With the growing importance of the ACSI model, we must determine which SEM techniques can better measure a CSI model. Cassel et al. (2000) suggested two criteria for choosing an SEM technique for a CSI model: (1) the SEM technique should determine the CSI score to make the comparison possible and (2) the SEM technique should exhibit good robustness of statistical properties.

The ACSI methodology has four basic features:

1. It uses an econometric model with measures of satisfaction and related constructs that are general enough to be comparable across firms, industries and sectors. These measures arise from survey responses. The relationship in the model and the variable measures used to estimate these relationships apply equally well to public services and competitive product markets.
2. The ACSI is itself embedded in a system of cause-and-effect relationships. If the model predictions are supported by expected behaviours of the constructs, then the validity of the ACSI is supported.
3. Satisfaction is measured as a latent variable using multiple indicators. Any one concrete measure like a single survey question rating is at best a proxy for latent satisfaction. Instead, the ACSI uses a variety of proxies or benchmarks that customers use to evaluate their overall consumption experience. These

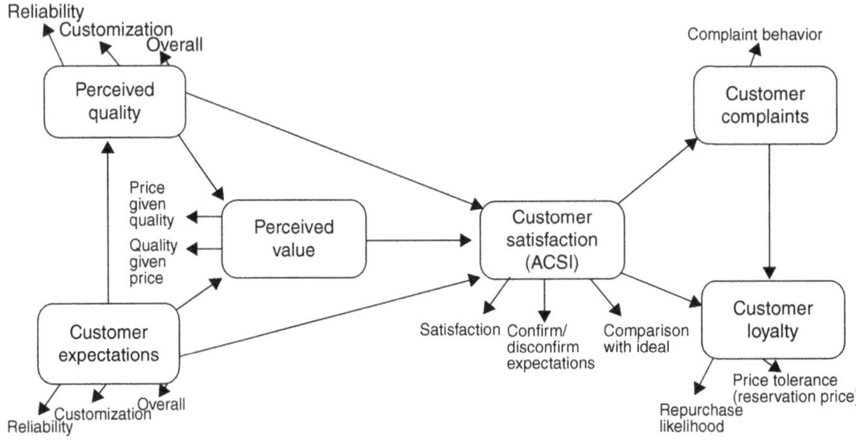

FIGURE 3.1 ASCI model.

proxies are combined into an index to operationalize satisfaction as a latent construct.

4. The primary objective in estimating the model is to explain customer loyalty, a construct of universal importance in the evaluation of current and future business performance.

The expanded ACSI model produces the latent variable Perceived Overall Quality as the composite of two latent variables: perceived product quality and perceived service quality. Exogenous variables in the model are customer expectations (which are influenced by the image of the organization created by its market promotion policies and practices, customers' experiences with other organizations and customers' specific likes and dislikes) and perceived overall value that takes into account price, given quality and quality given the price. Endogenously developed variables are customer satisfaction, customer loyalty and its opposite customer complaints. Thus the model is one of structural relations that are usually solved by the method of partial least squares.

The model can be written as a series of equations to be solved by partial least squares and can be formulated as:

$$E\,[\eta \mid \eta, \zeta] = B\,\eta + \Gamma\,\zeta$$

where η and ζ are vectors of unobserved endogenous and exogenous variables, respectively.

B (m × m) is a matrix of coefficients (to be estimated) for η and Γ is the corresponding matrix of order m x n for ζ. This implies that $E(\eta\,\zeta') = E(\zeta\,\zeta') = E(\zeta) = 0$, where

$$\zeta = \eta - E\,[\eta \mid \eta, \zeta]$$

The general equation relating the latent variables with the empirical (manifest) variables is $Y = \Lambda_y + \varepsilon$ and $X = \Lambda_x + \delta$

Here Y and X are the vectors of measured (in terms of survey results) endogenous and exogenous variables, respectively, while Λ_y and Λ_x are the vectors of relevant regression coefficients. Elements in the X and Y vectors (in case of all sectors except manufacturing / durables) are as follows:

X_1 = customer expectation about over-all quality
X_2 = customer expectation about reliability
X_3 = customer expectation about customization.
and
y_1 = over-all quality
y_2 = reliability
y_3 = customization
y_4 = price, given quality
y_5 = quality, given the price
y_6 =overall satisfaction
y_7 = confirmation of expectation

y_8 = distance from ideal product (service)
y_9 = formal or informal behaviour
y_{10} = repurchase intention
y_{11} = price tolerance (reservation price)

$$ACSI = [E(\zeta) - \min(\zeta)] / [\max(\zeta) - \min(\zeta)]$$

Here $\min(\zeta) = \Sigma\, w_I \min(\times I)$ where ζ is the latent variable for customer satisfaction, E stands for expectation over responding individuals, x_i's are the measurement variables of the latent customer satisfaction in terms of the responses to survey questions and w_i's are the weights of these variables. In ACSI, customer satisfaction indicators range from 1 to 10.

3.9.2 Employee Engagement Index

Employee engagement is an essential pre-requisite for excellent organizational performance. While customer satisfaction is both a driver of organizational excellence (even in situations where customers are potential beneficiaries or people mandated to be served) and a result of excellent performance, employee engagement goes beyond employee satisfaction, which is a multidimensional construct. Employees have to be satisfied with their jobs, should perceive matching their knowledge and skill levels or be motivated to augment their existing knowledge and skill sets, not affecting their physical health adversely nor conflicting with their family lives. They have to feel satisfied with the workplace environment in terms of both physical facilities and conveniences being up to their expectations and the relationship with leaders, peers and juniors good. They must feel satisfied with the compensation they receive by way of wages, salaries and perquisites and with career growth opportunities. These facets of satisfaction are collectively reflected in employees feeling proud of their identification with the organization and their desired continuity with the organization and, more than that, displaying commitment and dedication in discharging their tasks and responsibilities.

An important facet of employee satisfaction that is influenced by leadership of the organization is how employees perceive their leaders. While such perceptions are not easily spelt out in concrete terms, leaders are called upon to (1) clearly spell out what management expects from each one of the employees in regard to the tasks and responsibilities assigned along with corresponding targets, (2) provide employees with necessary advisories, facilities and guidelines, (3) empower them to feel that they are owners of their respective jobs and tasks and corresponding enabling resources and (4) ensure that employees are aligned with organizational values and culture and they are manifestly committed to discharging their responsibilities with dedication and devotion, at the same time taking care to make employees feel that they are not being overused but are only being helped to develop their potential.

An index is computed by using the Likert scale or some other similar scale to convert responses in a survey questionnaire on different aspects of customer satisfaction as is perceived by the responding employee and the manifest aspect of their performance in the organization. Usually, responses by way of 'fully agree, agree, neutral, disagree, strongly disagree' are sought against positive statements about manifest conditions and latent perceptions. The scaled scores are normalized to lie between 0 and 10 (or 100) and the weights are usually calibrated mostly be the relative importance attached by the organization to the aspects of employee engagement covered by the questionnaire.

3.9.3 Some Other Measures

Impact on the environment is not generally assessed in terms of an index. However, a measure of this impact by way of a determinant (if not a driver) of organizational excellence can be obtained as a check of conformity of the organizational activities with provisions of prevalent environmental norms. Measures of discharge into atmosphere, neighbouring land and water bodies and adverse effects on local flora and fauna as well as human habitation and their means of livelihood (especially those derived from land and water resources) can be worked out in specific cases. Currently, pollution control authorities operate in most countries and environmental norms cannot be flouted without attracting penalties.

Of late, society has been quite conscious of what it contributes to the growth and sustenance of organizations it supports and about the obligations of the organizations to the society. In fact, society has been demanding greater and more active roles of organizations in helping local communities to access better educational, healthcare and employment opportunities. Some authors have spelt out what they call the Society Trust Index. Once again, a carefully crafted questionnaire has to be designed, backed up by some forms filled up with relevant figures as are available with concerned official agencies, for the purpose of developing such an index.

4 Measuring Performance in Future

4.1 FUTURE OF MEASUREMENTS

While the dictum of Peter Drucker "What gets measured, gets done" was being well received, maybe with a pinch of salt in some cases, the need for a new direction for measurements of performance was being realized towards the close of the twentieth century by many established organizations aiming to record further growth or excellence. In fact, it was being felt that the indices (single-figure summaries) reflecting the effectiveness and/or efficiency in performance in an important function, or key performance indicators or figures of merit and metrics based on data collected for a reference period would generally provide a picture—sometimes quite comprehensive—of the current situation (applicable to the latest reference period). These metrics may be presented relative to what had happened in the past or what was the figures for the latest reference period in a few benchmark sets. Those indicators would usually ignore causal connections among themselves to explain the recent situation and predictions about what is likely to occur in future. Another realization that was gaining ground was related non-financial indicators that are likely to affect business performance in the long run. A concern for sustainability was growing and management or leadership across the board felt "Don't just measure what you can measure, measure what matters" and 'what matters' do relate to the society and the environment, beyond of course the people.

The need for consistency and sustainability of performance as defining characteristics of the long march to organizational excellence need not be overemphasized. Organizations with a proven track record of impressive performance during one period could not hold on to their positions of strength and supremacy and even were getting out of the race. The problem was not with measurements though there were deficiencies in data, models and data analysis used for some measures. It was more about measurements not revealing much about changes taking place constantly or about coordination and collaboration among inter-related functions and teams or about impending adverse impacts of happenings inside the organization and, more so, in the external world encapsulating the markets, the investment climate, the regulatory framework and the society.

Some of the noticeable changes in the development and deployment of performance measures over the last three or four decades are the following:

DOI: 10.1201/9781003591733-4

- A shift towards non-financial metrics, focusing on social, environmental and economic aspects of business.
- Emphasis on real-time measurements, with data analytics providing performance measures at frequent intervals as fresh data are accessed.
- Attempts to customize measures, suited to the specifics of the organization and providing the required extent of privacy and security.
- Integration of various analytics, machine learning, computer vision and Artificial Intelligence for faster, finer and more insightful information.
- Collaboration and integration of measurement systems across different departments and functions to promote aligned actions.
- Enhanced focus on the use of predictive analytics to come up with timely predictions of scenarios likely to develop and which may impact business performance adversely, not merely in terms of time-series analyses of performance measures generated internally, but by capturing relevant data adequately to link the performance measures with external factors and their temporal behaviours.

Technology is increasingly used and likely to be upgraded in different functional areas, covering both process technologies as well as digital technologies used for other functions, besides adequate consideration of the organization's priorities. These and other related developments have led to the emergence of several frameworks for measuring organizational performance. Two of them not mentioned as frameworks are the balanced score card introduced by Norton and Kaplan in 1992 and the triple bottom line introduced by Elkington in 1994. The Extended Performance Satisfaction Index (EPSI) (viewed by some experts as an extension of the European Customer Satisfaction Index ECSI) is a parallel case, which, however, comes up with a single figure or index, unlike the previously mentioned two. Moreover, EPSI use has been restricted to some European countries, while the balanced score card as a comprehensive analytical tool and the triple bottom line as a path-breaking concept have been accepted widely across the world.

Much before the concept of corporate social responsibility was formalized and incorporated into public policies related to business houses, some leading industrialists and businessmen started investing in efforts to improve education, health and infrastructure facilities available to the people and were even deriving business benefits from such investments. In fact, as early as 1953, Brown remarked that business has the obligation "to pursue those policies, make those decisions and follow those lines of action which are desirable in terms of the objectives and values of our society". Subsequently, concepts of corporate responsiveness and corporate citizenship were developed and a model to assess corporate social performance emerged. Corporate social performance can indeed be looked upon as one step ahead of the triple bottom line. Currently, many organizations have initiated measures to assess their corporate social performance and to link such measures with those of financial performance.

The future of performance measurement will be guided by a huge amount of data gathered from various sources and through the use of Internet of Things within the organization as well as other data automatically captured and fed into the cloud or data centre, and through computer vision, deep learning and generative Artificial Intelligence. Artificial Intelligence may be deployed to sift through the continuous flow of data-based measures of performance—organization-wide or function-specific—and to empower strategic leadership to take appropriate decisions and provide adequate directions for actions aimed at improvement for implementation. Looking beyond, AI can itself be used to generate actionable decisions for onward transmission to concerned functionaries, keeping strategic leadership in the loop. While the above provides a glimpse of what may happen in organizations with necessary digital resources and human capabilities, available or likely to be available shortly in some organizations, unplanned and even irresponsible use of these resources without adequate preparations may also cause damage to business prospects.

4.2 BALANCED SCORE CARD

History is replete with instances where initiatives for improvement in different spheres of human activity had their roots in the introspection of failures to achieve the set goals and targets. And this is exactly what characterizes the development of the Balanced Score Card and its adoption by industries in the United States during the 1990s, followed by its review and reports on its context-specific elaboration and implementation by researchers and professionals continuing till date. It all started from the findings of a research project that revealed the following:

- 90 per cent of companies fail to execute strategies.
- 85 per cent of the executives spend less than one hour per month in discussing strategy.
- 60 per cent of companies do not link budgets to strategy.
- Only 25 per cent of managers have incentives linked to strategy.
- Less than 5 per cent of the workforce understand the organizational strategy.

The need for a strategy-focused approach, involving a method of communication of the strategy across the organization and an alignment of the entire workforce with the strategy, along with the design and performance of processes aligned to the strategic objectives, was strongly felt by Kaplan and Norton who introduced the Balanced Score Card. They proposed it as a framework to be used by industry leaders and top executives to orient organizational performance towards the goals and objectives laid down in the strategy. Given the underlying assumption that the strategy has been developed as a concrete actionable version of the vision and mission of the organization and that the latter stress on sustainable growth and continuous improvement, the score card is meant to promote organizational performance in the firm's journey towards excellence. As pointed out by Kaplan and Norton, the score card is not to be regarded as a traditional control system but should be regarded as a communication, information and learning system.

The four perspectives in the score card (as indicated in Figure 4.1) go beyond internal processes and results of process performance, as they focus on (a) how customers (in its widest possible connotation) view the organization and how the organization in terms of its interfacing activities likes to be viewed by the customers and (b) how does the organization learn from its past and current successes as well as failures, how can it reduce and even eliminate failures and cash upon its potentials for

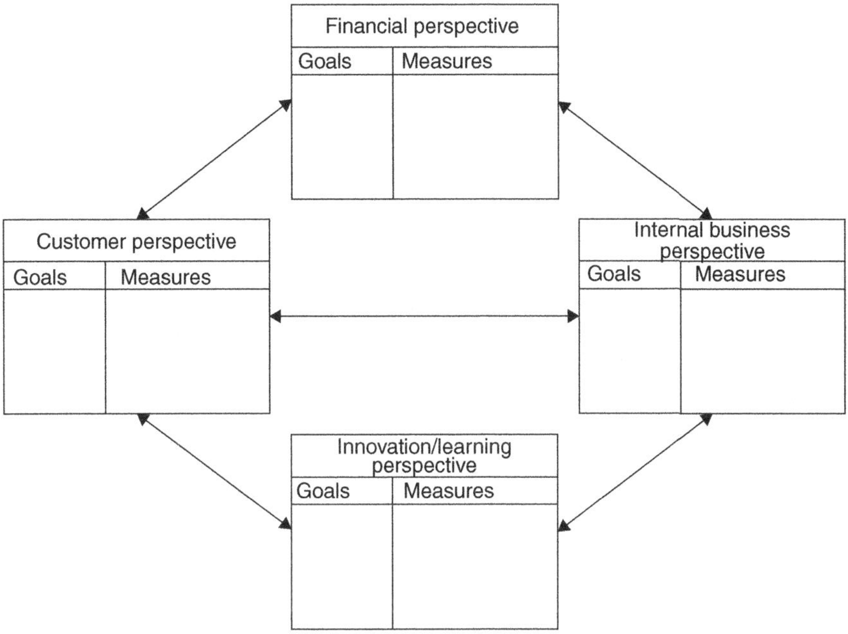

FIGURE 4.1 Balanced Score Card.

success. The results deal mostly with financial aspects of costs, revenues and profits as well as related expectations of the customers, the investors and the organization itself. For all the above reasons, the score card 'balances' financial with non-financial results, short-term achievements with long-term potentials, hard objective measures with soft subjective assessments, and internal business processes with external needs and constraints.

Kaplan and Norton suggested four to seven measures of performance from each perspective, some of which could be generic and some others suited to a given context. In the financial side, three themes have been taken note of: (1) revenue growth and mix; (2) cost reduction and productivity improvement and (3) asset utilization and investment strategy.. The first theme emphasizes expansion of product and service offerings, reaching new markets and new customers, switching over to higher-value-adding product and service mix, price rationalization, focus on customer profitability and the like. Reducing direct costs of production, avoidance of costs of poor quality, sharing common resources with other business units or verticalsamong other things. The third theme requires just sufficient deployment of productive assets, effective maintenance of assets and reduction of working capital level to the minimum required. There is no need to have as many measures or indicators as the number of issues needing attention.

In the perspective related to customers, we should keep two distinct sets of issues to define suitable measures. In the first, we focus on customer satisfaction and the resultant customer retention, besides customers' needs and expectations from new product and service offerings. As expected, the relevant measures here are based largely on latent variables such as perceptions and subjective assessments. The second set comprises identification of new markets and new customers, customer profitability, market share in the existing and the potential segments, and the like. Measures derived from customer surveys, despite being widely claimed by market research organizations as being capable of providing credible inputs for market-related decision-making, are quite likely to be fallible owing to a variety of reasons. A brief on the derivation of a customer satisfaction index has been attempted here.

The present author has a strong feeling that the second set of measures or indicators to answer the question, 'How does the organization present itself to the customer?' should include a critical review of the possible impacts of its market promotion policy and particularly the advertising policies and practices. Tall and unwarranted claims about quality, shelf-life, durability, effectiveness and other performance parameters of products being offered, put up in advertising media, often raise customers' expectations, which are quite often belied in real-time purchase experiences, causing harm to the market prospects of those products and services. And with customers getting quick and convenient access to requisite information, it should be natural to expect that customers will be able to detect any misinformation being shared. It is better to remain modest in terms of actuals than to heighten expectations, and then causing dissatisfaction, which has the potential to harm business prospects. To incorporate customer profitability, the customer profile needs to be revisited, and the focus be placed on profit-yielding customers and gradually withdrawing from non-profitable customers, even if they are old and faithful. So we require an activity-based costing system to work out expenses to serve customers and the revenues earned for each distinguishable group of customers. Kaplan and Norton draw attention to three attributes bearing on customer value propositions: (a) product/service features and the organization's role in updating and improving such features; (b) customer relationship especially in the area of service delivery and in responding to customer suggestions and complaints and (c) image of the organization and its reputation as a trustworthy and responsive supplier of products and services.

As regards internal business processes, both core and support processes must be aligned to customers' needs and expectations, on the one hand, and to stakeholder expectations in terms of financial results and stakeholder benefits arising therefrom. The role of innovation in designing new processes and novel products and appealing services is crucial to derive long-term value. The short-term value chain has to be definitely paid attention so that productivity increases and costs are duly controlled. The Plan-Do-Check-Act method suggested by Deming can be applied, aided by the use of SIPOC (Supplier, Input, Process, Output, Customer) analysis as a tool. A comprehensive review of past and current performance of the organization in the critical processes may indicate where the organization is relatively strong and advantage can be derived by focusing on such processes to modify the existing product/service profile.

In order that the score card can be used to look ahead and assess future prospects of growth and excellence, the fourth perspective on learning and growth is a significant addition to the conventional list of performance indicators. In fact, whenever a breakthrough improvement as against a marginal one is desired, existing capabilities of people, procedures and systems have to be adequately enhanced to embrace technological developments, radical changes in human behaviour and customer expectations and emergence of person-free generic procedures and self-correcting systems. Needless to say, breakthrough improvements are most often needed to guide organizations on their journeys to excellence.

While the score card with several well-knit components provides a comprehensive picture of an organization's performance, judged from different angles and not focused on financial results only, the processes considered are mostly internal (to the organization), though the customer perspective takes into consideration the external world to some extent. The customers extend to a wide array of stakeholders such as investors, regulatory agencies, business partners and even public service providers. The score card does not explicitly or directly address any societal or environmental concerns. But the score card fails to reflect the growing concern about the impacts of various activities, especially those related to disposal of wastes and other discarded materials and stores performed by organizations on public health and environmental quality. Considerations of sustainability of natural resources on which operations of business organizations depend do not find much of a reference in the score card. This leads us to discuss the triple bottom line concept and its implementation in the next section.

4.3 TRIPLE BOTTOM LINE

Nearly four decades ago, global realization of nature, or better, environment being both the most important source of all input resources for human activities and the ultimate sink for all wastes created by anthropogenic activities led world leaders, including leaders of industry, business and trade, to give serious thoughts to sustainability. The publication of the report "Our Common Future" by the World Commission (also called the Brundtland Commission) on Environment and Development in 1987 led to many discussions, debates and decisions regarding depletion of natural resources, deterioration in public health and impact thereof on human productivity and inter-generational equity. It was in this background that a famous entrepreneur and author John Elkington, working for the thinktank Sustainability, came up with the phrase 'triple bottom line' in 1994, marking a significant departure from the traditional business accounting system that focus on a single bottom line, that is, profit (maybe in terms of several related measures) as the only measure of organizational performance. The triple bottom line reporting of an organization evaluates performance of the organization in three key areas: people, planet and profit. This marked a new era in business performance measurement and reporting, which has eventually resulted in voluntary as well as regulation-mandated activities of an organization to improve quality of life, at least in its neighbourhood, if not generally. It became gradually known that a forward-looking, well-meaning organization can earn adequate profits and, at the same time, make the planet Earth a better place to live and work. To start with, the oil company Shell incorporated triple bottom line in its first Sustainability Report in 1997. With growing interest in the concept and its inclusion in business accounting practices, a transparent and consistent measurement framework has resulted in the creation of the Sustainability Accounting Standards Board as a non-profit body to promote the use of triple bottom line. Incidentally, the three P's have been sometimes taken as people, planet and prosperity. On a somewhat disparate note, one recalls Elkington saying, "The triple bottom line wasn't designed to be just an accounting tool. It was supposed to provoke deeper thinking about capitalism and its future."

Through his many writings, Elkington himself provides thought-provoking explanations for the origins of triple bottom line, its drivers, its distinctive aspects to be taken care of by top executives as well as a strategy for its implementation along with the benefits and shortcomings. Elkington refers to significant paradigm shifts which justify a recognition of revolutions in seven key areas or aspects as drivers of triple bottom line or as prime motivators to think about people and planet. These are the following:

1. Markets where the old paradigm of 'compliance' has yielded place to 'competition' as the new paradigm
2. Rigidly followed organizational 'values' have been replaced by 'soft' ones
3. Transparency in operations and in results were 'closed' within specified groups and now become quite 'open to all' concerned.
4. Earlier lifecycle technologies were meant for products now incorporated within functions
5. In terms of practice, partnership earlier was based primarily on compulsion for the time being and accommodated subversion, not being based on mutual trust; now partnership implies symbiosis to foster mutual success.
6. A different concept about 'time' has evolved in some sense. The earlier paradigm of looking at a wider canvas, maybe in the near future only, has been taken over by a bid to look at the existing canvas, but in terms of a longer time horizon and
7. Corporate governance practiced earlier was 'exclusive' in the sense that it was confined to the corporate board room and was concerned only with executives rather than the workforce as a whole or only with internal processes with the exclusion of their environmental impacts. Now, corporate governance is much more 'inclusive'.

Triple bottom line is all about sustainability of performance (encompassing both processes carried out and results achieved) and is an essential concern for progress towards excellence. Three aspects of sustainability correspond to the three bottom lines in the triple bottom line reporting framework. Social sustainability defines the first P (people), environmental sustainability corresponds to the second P (planet earth) while the third P (profit or economic prosperity) stands for economic sustainability. As is quite expected, each aspect of sustainability should be reported in terms of a suitable number of metrics. Metrics used within any P may be expressed in different units, may not be convertible meaningfully to monetary equivalents and hence may not pave way for aggregation to yield a logically interpretable index or figure of merit. This problem is more acute while dealing with metrics used for the three different facets of sustainability. The oft-prompted idea to monetize any metric (not expressed in monetary units) may not be philosophically or ethically acceptable. For example, the threat to or loss of some endangered plant or animal species arising from the impact of operations carried out within the organization or transportation to and from the organization of outputs and inputs or wastes discharged into waterbodies or pasture lands or other ecosystems cannot and should not be expressed in dollars. Thus the generally agreed view is to report a metric on its own unit relative to the goal or target to be set by the top leadership and duly incorporated in the corporate strategy. Even in the area of healthcare of its own employees, an organization should have a target for improvement in physical and mental health of its employees as can be quantified through necessary clinical and observational tests and not in terms of expenditure to be incurred on that account. Several experts have advocated the use of multi-criteria decision analysis tools like analytic hierarchy process to combine different metrics within any facet of sustainability and for the more ambitious purpose of reporting the overall performance on sustainability issues.

Metrics which are commonly deployed in this context to reflect on environmental, social and economic sustainability are illustrated briefly as follows:

- Environmental amount and value of renewable energy generated and/or used (during a reference period or on an average per specified period)
- Amount and value of materials recycled
- Volume of water used and drawn from neighbouring water bodies
- Amount of wastes released after treatment and (without treatment) into atmosphere, soil and water bodies in the neighbourhood
- Amount of greenhouse gases and other toxic substances emitted
- Number of trees planted, waterbodies cleaned, shrubs and bushes cleared, within the premises of the organization and outside
- Extent of deployment of devices to arrest noise and dust pollution and the like
- Social employee turnover, for reasons associated with poor pay and allowances, temporary nature of employment, poor promotion prospects, regularity in payment of salaries and wages, frequency of payment of bonus and special allowances, conveniences provided to workers and/or their family members by way of subsidies in accommodation, transport, education etc.
- Percentages of contractual and of casual workers in the workforce
- Number of regular jobs created during the last reference period
- Extent of adequate illumination, space for free and safe movement, toilet facilities and the like
- Economic realization of customer payments in time
- Timely payment of dues to suppliers and other business partners
- Return on assets (RoA) by considering their depreciated/imputed values
- Total taxes paid, including tax deduction at source for payments made to external parties with a tax deduction and deposit obligation
- Total subsidies received, total value of exemptions allowed and enjoyed
- Ratio between administrative costs and manufacturing/service production costs and the like.

As mentioned, these are meant just to illustrate the variety of metrics in use. It must be noted that organization-specific metrics play an important role in revealing the performance of an organization and those have to be identified and included by corporate management.

The second P standing for people is an umbrella term covering internal members or employees at all levels, besides customers, regulators, and shareholders. The most important component of people is workforce and organizational performance depends to a great extent on their contribution that, in turn, is affected by how are they treated by the top leadership. The latter should be evidenced from efforts and expenditures made by the organization in leading, supporting and developing people. In fact, these three dimensions have been elaborated in terms of 30 indicators used to assess the performance of an organization in the framework for Investors in People Award.

4.4 CORPORATE SOCIAL PERFORMANCE

Corporate social performance has now been recognized as a vital component of organizational performance, especially in the case of organizations whose performance is above the 'survival' mark and who would like to move further. A complex construct that evades a clear operational definition, this implies an integration of corporate social responsibility, corporate social responsiveness and responses to issues through an approach based on principles and policy. While social responsibility is an obligation or a responsibility, responsiveness is a trait of the organization that is reflected in the way the organization can and does respond to social and environmental issues, even proactively on the basis of anticipation. And beyond responsiveness, responses to social and environmental issues by way of actions by the organization and their outcomes are more important in determining the corporate social performance.

Table 4.1 by Wartick and Cochran (1985) notes the points of difference between CSR and corporate social responsiveness.

The approach adopted by Wartick and Cochran to assess CSP was extended by Wood (1991) as "a business organization's principles of social responsibility, processes of social responsiveness and policies, programmes and observable outcomes as the relate to the firm's social relationships".

A well-accepted three-dimensional model for corporate social performance suggested by Carroll (1979) recognizes philosophy of social responsiveness, social responsibility and societal issues involved (in terms of actions and outcomes thereof) as the three dimensions of CSP and the categories on each have been taken as pro-action, accommodation, defence and reaction for the first dimension; discretionary

TABLE 4.1
Social responsibility vs Social responsiveness

	Social responsibility	Social responsiveness
Major consideration	Ethical	Pragmatic
Unit of analysis	Society	The Firm
Focus	Ends	Means
Purpose	Window out	Window in
Emphasis	Obligations	Responses
Role of the firm	Moral agent	Producer of goods & services
Decision framework	Long-term	Medium and short term

(philanthropic), ethical, legal and economic for the second and consumerism, environment, discrimination, product safety, occupational safety and shareholder for the third dimension. While the model is based on sound logic, using it to measure CSP remains a problem and has been attempted only through a somewhat subjective approach based on responses in a survey of individuals and groups likely to be affected or covered by the actions taken by the organization. Responses may be converted to scores by adopting Likert scale, ranks could be assigned on the basis of scores and these could be aggregated to yield some measure of CSP.

4.5 DATA, MODELS AND DATA ANALYTICS

Data and data analytics have come to be largely applied in organizations across the spectrum to analyse the current health of the organization, to identify potential contributors to problems currently hindering growth and success, to predict what would happen in future if some changes are introduced, to work out optimum set-ups of different processes and to improve decision-making to accelerate the journey towards excellence. In this context, one should not forget the situation prevailing in industry and business a few decades back, when more data were collected than could be processed. More information was generated than was used as inputs for decision-making and more decisions were taken than were translated into action. A careful and planned avoidance of such a situation is fundamental to a meaningful adoption of recent advances in information technology and artificial intelligence along with associated developments.

Measures of organizational performance should reveal the current health of an organization in terms of various parameters related to resources or inputs, processes performed and outputs or outcomes achieved. An analysis of any such measure or a sub-set of related measures reflecting on the same aspect of business may shed light on the pattern or trend in the given aspect that may guide proper remedial or improvement decisions and actions. The leader should know the likely consequences of introducing some changes in a process on the ultimate performance of the organization before deciding on the nature and extent of change and its coverage. Even beyond introducing a few changes for improvement in process(es)–core or support—leadership has also to take effective decisions to manage and sustain the change.

These distinguishable purposes of using data analytics, machine learning and AI tools have been recognized as descriptive analytics, diagnostic analytics, predictive analytics, prescriptive analytics and cognitive analytics. From simple visualization tools to complex multi-layer artificial neural networks, from tracking the purchase behaviour of customers to use of support vector machines for classifying customers into two groups (one to be retained in terms of a plan and the other left to itself) based on the recorded purchase behaviour, from optimal setting of process parameters based on factor-response relations to acting on signals about equipment condition to initiate maintenance action, from faithfully recording and analysing responses in exit meetings to freeing up employees at higher skill levels from routine jobs and encouraging them to think and act creatively, and many more such benign applications are fast becoming essential and not just desirable.

Since a reasonably good idea about organizational performance is an absolute necessity to start with for any organization willing to undertake the arduous journey to excellence, let us focus here on how data, models and data analytics can be and are being used to develop objective, data-based and future-oriented measures of performance. In this section, we like to briefly consider two areas of performance: customer experience and people engagement. In the specific context of organizational excellence, we will focus on how an organization responds to feedback from customers—collected by the organization from customers or compiled from other sources—and how the organization deploys its people, motivates them, retains them and derives the best from their knowledge, skill and creative ideas. In the first case, customer complaints received may be analysed to identify and initiate steps to prevent their recurrence in future, customer suggestions and advice be promptly acknowledged and quickly examined to take decisions that can be communicated to the customers concerned at the earliest and decisions and actions regarding products and services in the offing be conveyed to existing and potential customers. In the second case, emphasis can be laid on knowledge management and, especially talent management, maintaining a workplace environment conducive to workers' commitment and identification with the organization and analysis of exit surveys or discussions to unearth reasons for exit so that steps for retention can be worked out. This is particularly important in the case of talent management. It may be added that the basic idea behind developing an effective measure of organizational performance is to identify and remove weaknesses and to build on strengths, looking back at the past as well as looking into the future.

With regard to customer experiences about products purchased or services enjoyed, as are communicated directly to the organization through mails or conversations, opinions posted on the web or doing the rounds in customer groups, we can start with data mining to build up a distribution of customer complaints in different categories like (1) delayed delivery; (2) product failure during use (before or after the expiry of any warranty period); (3) user manual not provided; (4) complaint not acknowledged; (5) complaint not acted upon within a reasonable time period; (6) failure attributed to customer fault; (7) inadequate compensation paid in case product failure is acknowledged as a fault on the part of the supplier organization according to value of the product/service in question and/or frequency of complaints during a certain period. Machine intelligence can be taken advantage of in checking the veracity or genuineness of a complaint, and if found genuine, the information needs to be passed on to the concerned process owners. A similar exercise can be taken up to identify high-value customers and to give more attention to respond to their complaints and even suggestions carefully and promptly. Suggestions for improvement in products and services including design and development of new products provided or solicited from high-value customers should be brainstormed, with the help of ChatGPT if needed, and predictive analytics used to generate the flow of net revenues from such new product or service for a final decision on the subject. Whatever the decision eventually taken, the same should be carefully communicated to the customers who provided the suggestions. Data analytics can be gainfully applied to examine the effectiveness of different forms of publicity and corresponding modes of campaign on sales. It will not be out of place to mention that not too often we come across publicity materials that simply heighten customer expectations about the value from a product or service to the extent that during actual usage the same is belied and a negative impact on sales is likely. Sales volume and value from different customer groups exposed to different publicity materials reaching potential customers in different modes can be examined with real data or even with data simulated on the basis of customer behaviour noted in the past and anticipated for the future using AI. Expected purchasing behaviour of customers exposed to different forms and modes of publicity can well be simulated through applications of AI. Customers (with a minimum number of transactions during a certain specified period) can be grouped into several classes based on their express or even latent concerns for delivery, quality and quality assurance procedures, billing requirements, payment schedules and expected response to complaints and requests using unsupervised clustering algorithm. The clusters may demand different approaches to retain them or to expand business with them.

In the domain of people engagement and knowledge management, useful applications of AI-assisted models and methods is yet to be reported. Organizations, especially those engaged in academic and research activities or in developing new designs, products, processes, and services, have to identify requirements of advanced knowledge or high skills or creative capabilities (in short 'talent') for high level of performance by determining gaps in the present situation and adopt a two-pronged strategy of retaining the existing talent pool and filling in the talent gap in a profitable manner. Some organizations provide additional facilities to nurture talent, mostly by way of material benefits that work only upto a certain level. Little is done to orient the workplace environment to foster talent, facilitate experimentation, focus on innovations and motivate the entire workforce to accept facilities, which are differentials for organizational excellence. Quite a few organizations carry out what they call an exit meeting between an employee leaving the organization and a senior management executive, usually from the human resource management area. A ChatGPT may be inducted to engage the leaving employee in a conversation to reveal reasons behind his exit with a focus on the opportunity to work in an atmosphere conducive to creative thinking, innovative ideas and freedom from routine and monotonous jobs associated with production or maintenance functions. In fact, these functions should be automated quite effectively to free up human energy for more constructive and creative jobs. The rate of attrition, among a particular group of employees, can be foreseen through predictive analytics applied to data relating to their performance and engagement level in the past. The predicted rate can help succession planning and ensure a steady availability of the workforce with desired capabilities.

5 Business Excellence Models Revisited

5.1 'EXCELLENCE': THE ULTIMATE GOAL

'Excellence' does not lend itself to a clear definition, not to speak of a consensus measurement. But most individuals and organizations would like, and even, strive to achieve excellence. Excellence goes beyond effectiveness and efficiency but may not necessarily imply growth—howsoever we characterize the latter.

We know that organizations operate at three levels—some survive (or even struggle to survive), some grow (in terms of size, volume of transactions, product and service profiles, turnover, and so on), while only a few excel. It is just possible that an organization continues to remain within its pristine boundaries but tries to achieve excellence in its operations and end results and—in some cases—might have reached a certain level of excellence.

Excellence implies staying ahead of the pack, exceeding competitors in the considered area. Excellence implies standardization, when it is needed, but it mainly denotes positive differentiation. In fact, standardization applies to all, while differentiation can be made only by a few of those who can. Differentiation in organizational performance can lead to excellence; but excellence is where the ladder ends and not the ladder itself.

Organizational excellence is reflected in excellence in organizational performance, with the word performance having two clearly discernible facets: execution and results. Thus, organizational performance would be understood in terms of execution of different operations carried out by the organization as well the results expressed in various dimensions. Quite often, organizational excellence will be referred to as business excellence.

Basically, measurement in the context of business excellence really refers to measurement of business performance. The main focus of commercial organizations is typically to achieve excellent financial performance (including creating best possible value for their owners). Often, this focus is widened to include non-financial performance (including creating value to other stakeholders like employees, customers, business partners and society). In particular, non-financial performance and stakeholder value is expected to include strong predictors for future financial performance and shareholder value.

DOI: 10.1201/9781003591733-5

Thus, any exercise to achieve organizational excellence and to sustain it, once achieved, involves

1. generating holistic performance indices (HPIs)
2. monitoring performance in terms of these HPIs and
3. strategic resource deployment based on HPIs.

There have been many constructs of the HPIs that capture both financial and non-financial performance indicators or success factors. Besides measures of the overall performance, we should also involve measures of the performance in sectors which pertain to the employees, customers, business partners, including suppliers and society.

The balanced score card is one illustration, where the organization is required to balance four different dimensions. It should not be merely a multiple score card. Sectoral performance indicators include Customer Satisfaction Index, Employee Satisfaction Index, Expected Performance Rating Index, Reputation Excellence Index and Society Trust Index. These indices take into account certain 'manifest' variables and use statistical analysis using a structural equations model to derive measures for the outcome variables which are 'latent'.

In order that these different measures can faithfully and effectively reflect the functioning of an organization, these should have high information quality (relevance, reliability) and transparency, verifiability and comparability to yield what is referred to as an accountable business performance measurement (ABPM) system.

5.2 BUSINESS EXCELLENCE MODELS

Models are known to provide simple (which may become simplistic in some cases) representations of reality that facilitates a comprehension of the underlying phenomenon or system and, thereby, its analysis and improvement. Given this generic idea about models, organizational excellence models (mostly referred to as business excellence models in the literature) have been proposed based on thoughts and experiences around excellence. It may be borne in mind that although we come across quite a few definitions of organizational excellence and ways and means to move towards excellence—noting that treating 'excellence' as a well-defined 'state' of achievable performance may be a subdued pursuit for excellence which knows no end—some thinkers consider the definition and means as inseparable. We may recall the nine principles mentioned by Pilcher to define and 'achieve' excellence:

- Commit to excellence.
- Measure the important things.
- Build a culture around service.
- Develop leaders to develop people.
- Focus on employee engagement.
- Be accountable.
- Align behaviour with goals and values.

- Communicate at all levels.
- Recognize and reward success.

An alternative for a model is a philosophical advice of Halders (2023) who suggests "Think differently, Plan differently, Act differently". Too abstract to implement in practice, this suggestion requires a lot of elucidation of the qualification "differently" in a given context, since a general elucidation applicable to all situations may not be available. To think, plan and act differently demands the organization to understand what peer organizations have been doing and to know 'what they think' may be well, nigh, impossible and to know about their plans calls for a cooperative interaction. It may be easier to view their actions. Accessing reports available on the public domain or even on request has been a practice of acting 'differently'.

It goes without saying that a model based on a comprehensive understanding of organizational excellence should work as a guide for any organization to link its performance in terms of results to the way it manages its inputs and processes. In the words of Dahlgaard, who has extensively studied these models, any such model is meant to be a non-prescriptive framework, somewhat generic in nature so as to be applicable to different types of organizations.

Most models to explain organizational performance, which are often taken as business excellence models, refer to leadership, policy and strategy, resource management, people management and process management as the determinants or drivers of performance. The first two influences the remaining three. Further, it is leadership (in a broad sense) that decides the policy and strategy.

The policy and strategy (along with the operational tactics) developed by the leadership guides the other drivers of performance. On the results side, we generally take into account customer satisfaction quantified in some appropriate manner like using a Customer Satisfaction Index, satisfaction of other stakeholders such as investors, business partners and employees, financial results revealed in the balance sheet and summarized by ratios like return on investment, gross value added per worker, debt–equity ratio, cost of poor quality as a fraction of the total sales billed, impact on environment and impact on society.

Institutional performance should be assessed in terms of results like winning an award for quality or for business excellence, ensuring a lucrative earnings per share, low attrition rate among employees or eliciting adequate cooperation and support from business partners and enablers which appear in the above model like creation and maintenance of a supportive and congenial environment for work or acquisition and deployment of adequate resources made available to workers in time.

Analysis of business excellence will imply identification of the factors which contribute to (improving) excellence and certain results they yield, as well as certain connections between these two groups. This is what is attempted by the business excellence models.

It would appear from the simplistic analysis of organizational performance attempted long back by Ishikawa that organizational performance is eventually determined by people who have to be led and deploying resources according to sound strategies and policies to carry out certain processes that yield the desired results.

More recently, we have several business excellence models that speak of the drivers (determinants) and the results (outcomes) of excellence. Typically, a business excellence model uses the following:

1. Leadership impacts on
    a. Strategy and policy
    b. People management
    c. Resource management
2. These three factors influence process performance.
3. Process performance is reflected in
    a. People satisfaction
    b. Stakeholder satisfaction
    c. Impact on society
4. The last three elements as mentioned in (3) determine business results in terms of financial and economic indicators.

Thus the crux of business excellence lies in enhanced quality of performance in critical-to-success processes.

The journey towards excellence indirectly implies 'working differently from others' and not just following a carefully crafted path developed by collective wisdom embodied in a group (can be named differently) honestly and effectively. However, a broad framework that incorporates several approaches to enhancing operational performance from which an intending organization can choose the one that suits its vision, mission and strategy may be accepted with or without modifications. It in this spirit that an organization pursuing excellence can make good use of the framework of a well-known business excellence model, not the right criteria for assessment or for award. In fact, the Baldridge framework and the European Foundation for Quality Management (EFQM) framework have been revised in response to changes taking place in the world of technology for the production of goods and services and changes in the environment within which production systems operate. Considering manufacturing industry in particular, these frameworks have duly adopted the upcoming notion of Quality 4.0 to fit into the needs and expectations of Industry 4.0.

5.3 FIVE PILLARS OF ORGANIZATIONAL EXCELLENCE

Good is no longer good enough. To thrive in today's competitive environment and even to survive, you have to excel. To excel, an organization needs to focus on all parts of the organization, optimizing the use and effectiveness of all of its resources and ensure that results obtained (at the end of an accounting year or a planning horizon of several years) can satisfy all its stakeholders, including the environment and the society at large. We have to appreciate the concerns for all the three P's—people, planet and profit—and not just the last-named one.

The following five key pillars of organizational excellence have been identified through years of research into improvement approaches and practices in a wide array of organizations and management has to move forward on all the five simultaneously.

- Process management (focusing on both core and support processes)
- Project management (considering all ventures for expansion and consolidation)
- Change management (essential in all steps to move forward along untrodden paths)
- Knowledge management (encompassing knowledge acquisition, absorption and use) and
- Resource management (emphasizing optimum utilization of physical, material, technological and human resources without overstretching any of these).

Business excellence models owe their origin to concepts, methods and practices to boost quality of products and services in manufacturing industries, examining the best practices and attempting to develop frameworks which could be adopted by various organizations to improve their performance. The essence has been to link quality improvement in processes and products as well as in related services to business results. Initially, these frameworks were used to assess the performance of organizations against the framework and to encourage good performance through awards. The frameworks did touch upon, explicitly or overtly, the five pillars mentioned above. Models starting with the Malcolm Baldridge National Quality Award (MBNQA) have undergone revisions commensurate with changes in organizational development, business environment, market requirements and investment climate.

5.4 MBNQA MODEL

The MBNQA model can be regarded as the earliest framework to elevate an exemplary quality management practice to one that qualifies for a roadmap to excellence in performance. Several national quality award models were developed on the MBNQA platform and used, like the MBNQA itself, as organizational excellence models. Instituted by the US Congress in 1988, managed by the National Institute of Standards and Technology and administered by the American Society for Quality, the MBNQA is the highest presidential honour for performance excellence.

The model has undergone a few changes through revisions, though the basic structure continues to remain the same. The enablers and the results constitute the 'categories' or 'criteria' treated as constructs. Each category has a few 'items' or 'questions' on which information is sought from the organization. The results include both non-economic and economic aspects and a distinct focus is laid on employee engagement. Thus, this roadmap to excellence does take into account social responsibility, though 'impact on environment' does not appear as a distinct item in 'results'. Adequate emphasis has been placed on quality management in the area of 'process management'.

The MBNQA places greater emphasis on 'enablers' or ' drivers' of excellence, looks upon leadership as the prime mover and recognizes 'strategic planning' and 'customer and market focus' as being influenced by leadership and as influencing, in turn, 'human resources focus' and 'process management'. Initially, 19 items or questions were included, though in the 2021–22 version, there are 17, with each of the first six categories or criteria—(1) leadership, (2) strategy, (3) customers, (4) measurement, analysis and knowledge management, (5) workforce and (6) operations—covered in terms of two questions or items, while the criterion 'results' has been captured through five items. For the first six, answers require information on the key processes bearing on approach, deployment, learning and integration. For the five items relating to 'results', information is sought about levels maintained, trend, comparisons with industry leaders using benchmarking and integration.

The core values and concepts used to build the MBNQA model are the following:

a. Systems perspective
b. Visionary leadership
c. Customer-focused excellence
d. Agility and resilience
e. Organizational learning
f. Focus on success and innovation
g. Management by fact
h. Societal contribution
i. Ethics and transparency
j. Delivering value and results

The 2023–24 version of the MBNQA framework focuses on organizational agility, innovation and transformation, risk management and supply chain resilience, societal contribution and environmental sustainability, changing nature of work and workforce needs and diversity, equity and inclusion.

Quite in tune with changes (including disruptions) taking place all around us, this model is all-pervasive and quite demanding on organizational leadership to implement it fully. It emphasizes on performance excellence implying excellence in performance of both core and support business processes. A whole lot of emphasis has been placed on leadership and on innovation supported, promoted and sustained by leadership.

The excellence models in circulation currently suffer from some limitations, which have been pointed by several authors, including the present one. It does not work well to predict future performance nor provides a roadmap for sustained results. In fact, among some MBNQA winners, product quality was not outstanding or even excellent, some did not continue to derive satisfactory business results and some had even to apply for bankruptcy. One reason could be that this model does not provide a comprehensive and integrative perspective of organizational processes (Mallat-Parasat, 2015). As can be easily appreciated, business results and those features which enhance such results (both financial and non-financial) like customer satisfaction and market focus are influenced more by performance on core as well as supporting business processes than on resources (including hard technical ones and soft skills). Some critics also point out the relatively less importance accorded to process management including process improvement and use of recent developments both in the area of technology and in the domain of quantitative methods geared to support process improvement.

However, some authors argue that these criticisms have cropped up because of inadequate understanding of the model, aggravated by the lack of data pertaining to the award process due to some privacy requirement by National Institute of Standards (NIS).

5.5 EFQM MODEL

The EFQM model is based on the assumption that the satisfaction of the customers, the employees and a positive impact on society can be achieved by means of political and strategic leadership, a right employee management, an effective use of the resources available and an adequate definition of the processes, which finally would lead to excellence in the results. The model (as presented in Figure 5.1) takes into account internal factors mostly, except for the qualification 'political' of leadership. There are five enablers leading to three results related respectively to people, customers and society and to the 'key performance results'. The focus on customers has appeared as an enabler in MBNQA and in some other models. Going by the weights assigned to the nine criteria or constructs, this model ascribes equal weights to 'enablers' and 'results'. Among the enablers, processes account for 28 per cent of the total for enablers, followed by leadership with 10 per cent. Thus processes rightly get the highest priority and include sub-criteria like (a) systematic design and management of processes, making the best possible use of skills and capabilities of the workforce deftly transformed into organizational capabilities through leadership skills and competencies; (b) introduction of continuous process improvement, based on innovation, leading to (c) design and development of new products and services, (d) production, distribution and customer service of products and services, keeping in mind the value chain and (e) management and improvement of the relationship with customers, as part of an intangible resource.

On the 'results' side, customer results, encompassing more than just customer satisfaction, account for 40 per cent of the total, with key performance results (mostly financial) accorded 30 per cent weight, people results carrying 18 per cent and the

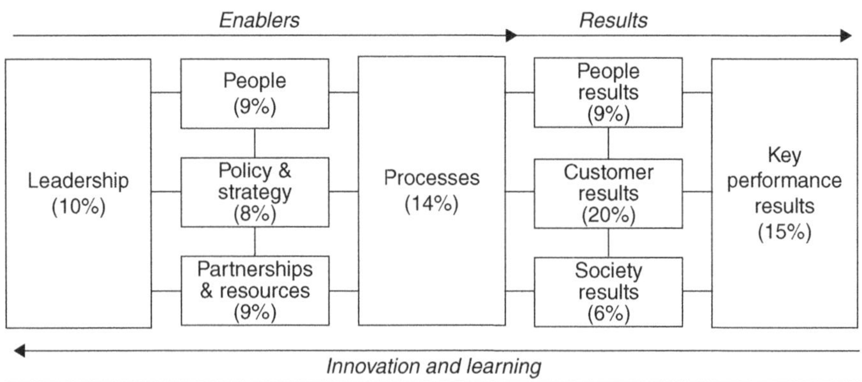

FIGURE 5.1 EFQM model.

remaining 12 per cent going to the somewhat difficult-to-measure society results. In a very broad sense, the EFQM model can be considered as a precursor to the later recognition that only profit does not constitute the bottom line of an organization's performance chart and that the three P's—planet, people and profit—together constitute the bottom line.

The current 2022 version of the EFQM model speaks of seven criteria comprising three segments—direction, execution and results—to answer the following three questions: Why does the organization exist? What purpose does it fulfil? Why has this particular strategy been adopted to fulfil that purpose? (direction)

- How does it intend to deliver its stated purpose with the strategy adopted? (Execution)
- What has the organization achieved till date? What does it intend to achieve in future and how? (Results)

The first segment comprises three criteria: purpose, vision and strategy. The second comprises engaging stakeholders including its workforce, creating sustainable value for the stakeholders and driving performance and transformation. Stakeholder perception and not simply figures on financial performance and strategic and operational performance are the two criteria within the third segment.

It is worthwhile to recall that initially the EFQM model had nine criteria, consisting of five 'drivers' or pre-requisites and four results. The model was based essentially on three pillars: people, processes and results. It is remarkable that in the latest version, there is no prescribed set of performance measures, though such measures have to be targeted and achieved by an organization trying out this model.

5.6 EXCELLENCE CANADA MODEL

Industry Canada introduced the Canada business excellence model in 1984 to promote performance excellence in the manufacturing sector. Later on, the National Quality Institute as an independent non-profit organization was operating this award and in 1992, it became Canada Excellence award with its coverage expanded to services also. The award does not involve any competition and the contenders are not ranked (as in the case of several other quality awards), Selection is based on meeting and even exceeding requirements laid down in the prescribed contemporary and holistic standards, developed for each category of awards depending on the type of activities and outputs of different organizations. Awards are presented to all the selected organizations within each of 11 categories. The excellence award is presented only for breakthrough results based primarily on the balanced score card.

A distinctive feature of the Excellence Canada award model is the explicit inclusion of workplace safety and metal health at work in the list of criteria. The recipients are role models in the areas of leadership, governance, strategy, planning, customer experience, employee engagement, innovation and wellness. Demonstration of continuous improvement and measurement of progress are among the basic requirements. Outstanding performance in the areas of innovation and financial wellness have been emphasized.

The award for excellence emphasizes improved strategic outcomes rather than outcomes of processes or results that can be achieved from improvement in tactical or operational strategy changes. These are

1. becoming more strategy focused and improving internal and external communication around strategy (embodying the organization's culture and value system),
2. alignment of people at all levels and both core and support processes around strategy,
3. effective use of performance measurement, analysis and review to improve fact-based decision-making reflected in improvement-oriented actions,
4. sustainable change in the organization's ability to transform and adapt to changing internal and external circumstances and quite expectedly,
5. significant improvement in financial results and stakeholder satisfaction.

5.7 A FEW OTHER MODELS

To help business organizations improve their performance and, incidentally, compete for awards in recognition of their efforts, quite a few models or frameworks have emerged at national and international levels. Possibly the earliest is the Deming Prize Award Model—not named as a Business Excellence Model (BEM)—instituted by the Union of Japanese Scientists and Engineers (JUSE) to promote applications of Total Quality Management, resulting in improved organizational performance, essentially in terms of financial metrics but focusing on both process management and people engagement. Almost contemporaneous with MBNQA, the Australian business excellence framework (ABEF) was developed through the efforts of the National Quality Foundation. The Brazilian model MEG was introduced in 1994. A somewhat different model applicable only to organizations within the same corporate group is the Tata business excellence model (TBEM). Started in 1987, currently operated by the standards body SAI Global, ABEF is an integrated leadership and management system based on 12 principles of excellence:

- Clear direction allows organizationl alignment and a focus on achievement of goals.
- Mutually agreed plans translate organizational directions into actions.
- Understanding of what customers value, now and into the future, influences organizational direction, strategy and action.
- To improve the outcome, improve the system and the associated processes.
- The potential of an organization is realized through its people's resourcefulness, enthusiasm and participation.
- Continual improvement and innovation depend on continual learning.
- All people work in a system, and outcomes are improved when all people work on a system.
- Effective use of facts, data and knowledge leads to improved decisions.
- All systems and processes exhibit variability, which impacts predictability and profitability.

- Organizations provide value to the community through their actions to ensure a clean, fair, safe and prosperous society.
- Sustainability is determined by an organization's ability to create and deliver value for all stakeholders.

Senior leadership's constant role modelling of these principles and their creation of a supportive environment to abide by these principles are necessary for the organization to reach its true potential.

The framework includes seven categories (of items) expanded into 22 items with both enablers and results within each category. The seven categories are: leadership, strategy and planning, knowledge and information, people, customer and market focus, innovation, quality and improvement, and success and sustainability. One item under leadership is environmental and community contribution. The results are assessed in terms of four dimensions: approach, deployment, results and improvement.

While the framework has a lot of similarity with EFQM as expected, certain items directly emphasized in ABEF, such as innovation, quality and improvement or, more importantly, sustainability, make it somewhat distinct. Similarly, the block of items on knowledge and information does not find a direct mention in some frameworks for business excellence. Of course, all the models are undergoing changes and converging more closely to a common entity. Introduced in 1994 and updated seven times till 2016, the Brazilian model deserves attention for including the relatively less emphasized items in the MEG. In fact, the model focuses on sustainable development through transformative leadership, compromise with stakeholders in situations of conflict, adaptability and system thinking, value generation, process orientation and organizational learning and innovation. With 39 questions covering these eight fundamentals, this model evaluated performance using a pathway involving four elements—focus on best practices, application, learning and integration of each factor. The results constitute the second-order factor while the other factors are recognized as first-order factors.

The Ibero American model for management excellence was initiated in 1999 by the Ibero American Foundation for Quality management and consists of five enablers: leadership and management style, strategy, people development, resources and associates, processes and clients. There are four items related to results: results on clients, on people development, and on society besides global results. The model stresses on developing alliances on innovation and sustainable future.

Not introduced as a BEM but to promote integration of TQM in a broad sense for improving organizational performance, the Deming Prize is awarded both to individuals as well as to organizations on the basis of an elaborate assessment process. The prize relates to a model based on Deming's 14 points of management and three main criteria: direction, activities and effects. Organizations that use TQM to achieve the following three criteria are awarded the Deming Prize.

The company has built proactive customer-oriented business objectives and plans based on social responsibility of the organization under clear management policies that reflect its management philosophy, industry, scale and environment. In their formulation, senior management demonstrates leadership. The company goals and strategies stated above are being realized with the appropriate use of TQM and as a result, the organizational capability necessary for future growth has been secured in addition to the achievement of the business objectives and strategies spelt out earlier.

5.8 SERVICE EXCELLENCE MODEL: CHARTER MARK

The Charter Mark was an award scheme launched in 1992 to recognize excellence in public services in the United Kingdom. It was an outcome of a political initiative called the 'Citizen's Charter' introduced by the then Prime Minister John Major. The system was reviewed in 2005 and was replaced by the 'Customer Service Excellence' standard in 2008. The new scheme is primarily focused on improving standards of public service provision and enhancing user satisfaction with public services.

The Customer Service Excellence standard is based on five criteria mentioned below:

- *Customer insight*: Effectively identifying customers, consulting them in a meaningful way and efficiently measuring the outcomes of service
- *The culture of the organization*: There must be a commitment to build and foster a customer-focused culture throughout the organization, from the strategic leader to the front-line staff.
- *Information and access*: Customers value accurate and comprehensive information that is delivered or available through the most appropriate channel for them. Organizations must strive for providing effective communications to customers.
- *Delivery*: Listening to the 'voice of the customers' about the service provided through comments, complaints and other feedback mechanisms can help an organization to make changes to the way the organization operates, which in turn, can improve service delivery.
- *Timeliness and quality of service*: The speed of response and meeting timelines are critical to customer satisfaction. Care must be taken not to achieve responsiveness at the cost of poor quality of service. A proper balance between the two must be achieved for best results.

The original Charter Mark scheme was a competitive award system. However, the new Customer Service Excellence scheme is not an award to win but rather a benchmark of service excellence that all service organizations should aspire to achieve.

5.9 SOME LIMITATIONS OF THE MODELS

The role of planning in respect of the determinants and targeting and predicting in respect of the results should be emphasized and an extended model incorporating the measures of effectiveness and productivity of the drivers taking into account data relating to the relevant past along with the current situation (during which period no major policy change took place) should be proposed. The business excellence models are generally used mostly for winning some award or at least gaining some confidence of being on the right track. They do not focus on sustainability and therefore on the sustenance of the ongoing efforts to leverage the determinants, drivers or key success factors. In the case of R&D organizations, progress in knowledge acquisition and knowledge management along with documentation and in-depth analysis of failures—major and even minor—in the past should be considered as an important driver of excellence. In fact, plans and programmes for knowledge acquisition through training and re-training, interactions with external experts besides in-house exchange of ideas and thoughts, representation of scientific workers in seminars, conferences and workshops, interactions with potential beneficiaries to elicit information about their expectations and needs and with funding agencies to know about their preferences and priorities, and similar other activities should constitute this driver of excellence.

In the applications of these models, weights are assigned to the different determinants and to the consequent results. This is done exogenously by considering the organization's value system or culture, making a comparison across organizations or across situations where the value system and/or culture does not remain the same. One suggestion could be to develop such weights endogenously by some method like data envelopment analysis. Weights for the drivers and the results justify separate computation.

The existing models identify enablers or determinants and consequences in terms of some suitable measure of organizational performance. They do not really link these two sets of variables (some manifest and some latent) explicitly in terms of qualitative relations or quantitative regressions. Indeed, application of structural equations modelling could be attempted in this connection. Further, the models do not directly consider the manner in which and the extent to which an organization harnesses or manages the determinants to achieve certain targets of performance.

Even with these deficiencies, business or organizational excellence models do provide the desired impetus to any organization for further progress and improvement in its intended journey towards excellence.

6 Leadership

6.1 SOME GENERALITIES

Leadership is by far the most important driver and enabler and, in fact, a pre-requisite for excellence. An organization without a leader is like a ship without a rudder and with each passing day, driven by rapid advances in science and technology, the sea is getting more and more rough and risky to navigate, even though the ship is getting more and more equipped to face turbulence. Organizations of all types are all in genuine need of effective leadership: an organization in the making yet to start operations, a newly started organization faced with teething problems of recruiting right people and procuring right resources, an established organization swamped by changes all around and struggling to continue operations, a vibrant organization performing pretty well finding emerging competitors strong enough to derail its performance, a start-up working hard to grow and achieve success and an organization sincerely trying to hold on to its current position in the face of unseemly eventualities..

Two categories of individuals associated with organizations of various types deserve recognition as leaders. In the first category are individuals who were involved in establishing the organization by providing funds or facilities, ideas and initiatives, and even carving out a vision for the future and developing a broad policy to be adopted by the organization throughout its journey. Also included are individuals who were not involved in any such capacity but were in the first group of workers discharging the more important responsibilities and laying out a path for future workers towards high levels of performance. Considering the success achieved by quite a few organizations, it may be mentioned that some of these individuals were highly accomplished, some highly skilled, some completely dedicated and passionate, some quite humble and that way were remarkable. At the same time, there were some who were not so communicative or persuasive, or even highly educated or skilled, though they could feel the needs of others, dream about apparently impossible events and embark on adventurous journeys. It is the second category of leaders who matter the most in the current scenario and for organizations, individuals who are capable of accepting responsibilities to guide and control different tasks and processes, left to themselves, are needed. Maybe, they may need a bit of advice, inspiration or direction from the promoters or founders. This defines what can called 'executive' leadership. Here we find several levels in a hierarchy, few in a flat organization but more in a

DOI: 10.1201/9781003591733-6

vertically hierarchical one. Executive leadership overlaps with management, essentially in terms of responsibilities.

We also come across organizations being led by individuals belonging to the first category or by their representatives (by way of relatives and other confidants) occupying positions of top-level executive leaders. And, in some cases, the latter have been able to ensure remarkable progress of their respective organizations. As expected, the possibility of a family-managed organization in the manufacturing or service sector is excluded for organizations in the public sector or for institutions with elected executives.

Given this context, let us discuss leaders and leadership along with some associated questions. Let us start with the proverb 'leaders are born, not made', which is refuted by the contention that leadership qualities and virtues can be cultivated and imbibed by individuals intending, or better, aspiring to be 'leaders'. In fact, process leadership theory is built around a process through which leadership qualities can be instilled in an individual, facilitated by proven leaders. This again is challenged by the argument that this intention or aspiration is not that common and reflects an inherent quality or feature of the individual, dormant or manifest. Do we really need to understand leaders as individuals and leadership as a responsibility accepted and discharged distinctly?

The next question we ask is 'Should we have only one leader?' Or, should there be leaders in every functional area, who, of course, should work in harmony with one another and with the 'top' leader, all being aligned with the overall vision and mission of the organization? The argument in favour of a single leader is that 'too many cooks spoil the broth' and given that communication channels are always noisy, communications from different leaders may be diluted or deviated. However, this should not become an insurmountable problem and leadership should be promoted and even ensured in all important areas of an organization's operations. More often than not, organizations need functional leaders who can direct, guide and support different broad functions, in tune with the goals and objectives set forth for the organization by top leadership and the strategy adopted to achieve the goals and objectives. 'Collective leadership' has been the order of the day and should be taken in the spirit of the adjective 'collective', which distinguishes between a collective noun and a noun of multitude in classical English grammar. We desire to have as many leaders as are deemed fit in a given situation unified by the principles and priorities accepted by the organization, and not just a multitude of leaders acting independently of one another in the absence of a unifying mechanism.

Just to illustrate, a university has a vice chancellor as its top leader who sets out the vision and mission for the university, if not already spelt out—keeping in mind the National Education Policy and the emerging needs of the society as well as the current trends in higher education and research in the international arena. While they combine both academic and administrative leadership roles, there will be usually be a leader for administration, a leader like the Dean to take care of academic activities, sometimes another for collaboration and outreach programmes, someone for research and development and the like who will be supporting the overarching role of the vice chancellor. In many universities, each faculty, such as science, humanities, law, engineering and technology, has a separate dean. Administration is usually taken care of by the university registrar and finance and accounts by another officer. These are executive leaders who are common members of university bodies like the Senate, the Syndicate or the Executive Committee to facilitate the process of aligning their roles with the goals, objectives and targets set before the entire university.

6.2 LEADERS VERSUS MANAGERS

A statement like 'we don't need managers, we need leaders' will hardly be disputed, except those who would argue that we need both. Managers are identified by the positions they occupy or the responsibilities they have to discharge within the authority they enjoy, while leaders are identified by the roles they play to help, support and guide their peers as well as subordinates and even superiors in resolving the 'problems' the organization faces.

In a somewhat lighter vein, the author remembers the following poem composed during the peak of the Japanese Quality Movement, titled "America No Kokoro (America's Heart)":

Let's get rid of management.
People don't want to be managed.
They want to be led.
Whoever heard of a world manager?
World leader, yes.
Educational leader, political leader, religious leader,
Scout leader, community leader, labour leader, business leader.
They lead.
They don't manage.
The carrot always wins over the stick.
Ask your horse.
You can lead your horse to water, but you cannot Manage him to drink.
If you want to manage somebody, manage yourself.
Do that well and you'll be ready to stop managing
And start leading.

The last three lines hint at 'leadership' as a higher ability than 'management'.

Management functions and leadership roles are differentiated and the need for both in any organization is justified. In this connection, it is relevant to consider the core of the numerous definitions of management like the one promoted by the International Organization for Standardization (ISO), which states that management consists in direction, guidance and control of an organization and all its activities. Focusing on the first and the third components, that is, direction from the beginning and control before or at the end, we can clearly see the need for management in its hierarchical manifestation in the form of chief general manager, general manager, manager, deputy manager, and so on. The role of leadership is really centred around the component of guidance. Guidance necessarily implies involvement—as against staying aloof—only initially, as and when required or sought for, or throughout. It is quite imaginable that the most desired situation is where managers are also leaders and all the three components are fully taken care of. In fact, direction and control includes assignment of tasks and responsibilities to those who have to be led and constitute the authority of management. Providing guidance is indeed a big and a critical responsibility.

Managers and leaders do not constitute a dichotomy and any organization needs both. Maybe we find both managers and leaders (in terms of designations or tasks and responsibilities) in some organizations and only managers in some others. After all, team or group leaders are also held accountable for the performance of their respective teams, thus bestowing upon leaders the managerial tasks of guiding and controlling activities of members and assessing their performance. To meet the desired performance target, they have to support the members and maintain a conducive workplace environment. The latter implies the leader's role to resolve interpersonal conflicts within the team, beyond ensuring safety and security and hygienic conditions in the workplace. Similarly, persons recognized as managers do display—implicitly or unknowingly in some situations—leadership qualities while discharging the responsibilities assigned to them by the executive leadership in the organization. In reality, the distinction between managers and leaders is becoming thinner by the day and we can well afford to ignore it. This will not confuse our understanding of a situation where the organization has managers to look after departments or branches or broadly defined functions on an ongoing basis alongside leaders to successfully execute some special tasks like introducing a new process and convincing the workforce in the corresponding area about its benefits. Bennis (1994) points out some differences between managers and leaders, which appear to be interesting, as provided in Table 6.1.

Let it be noted that notwithstanding differences between managers and leaders, in our discussions on leadership as the prime mover of an organization in its journey towards excellence, we will continue to use the term 'leader' broadly to include 'management' also.

TABLE 6.1
Manager vs Leader

Manager	Leader
Administers	Innovates
Is a copy	Is an original
Maintains	Develops
Focuses on systems and structure	Focuses on people
Has a short-range view	Has a long-range perspective
Asks how and when	Asks how
Has his eye on the bottom line	Has his eye on the horizon
Imitates	Has original thinking
Accepts the status quo	Challenges the status quo
Does things right	Does the right thing

6.3 FACETS OF LEADERSHIP

Leadership has been a subject of intense discussion, resulting in a multitude of leadership theories, styles, habits and practices that naturally reflect quite a fair amount of commonality and some distinguishing features. While leadership is the driving force for all that is planned, executed and evaluated by any organization and therefore the most important enabler of excellence, what needs to be considered in the context of organizational or even operational excellence is how leadership provides direction, guidance, support and control over the different processes carried out currently or to be introduced in future by any organization. As pointed out in the beginning, some organizations under consideration may be involved in delivering services, unlike some others engaged in designing and manufacturing physical goods. Even within manufacturing and, similarly, within services, the roles and responsibilities of leadership could vary from the public to the private sectors, from simple, FMCG (fast-moving consumer goods) to expensive, complex equipment manufacturer, from academic institutions to public health service providers, and so on. Expectations from leadership vary across the wide spectrum of organizations.

The volume of publications and materials available on the web dealing with leadership is simply bewildering. Being generally verbose, some books and articles stand out clearly as providing clear and concrete ideas about the three distinct—though interrelated—aspects of organizational leadership: (1) skills and abilities to be possessed and cultivated by leaders or by people to act as leaders, (2) traits or components of behaviour or habits as are exhibited by leaders consistently and (3) roles and responsibilities to be accepted and discharged by leaders, in both overt and covert ways. There can be levels of leadership in terms of the degrees of each of these three aspects possessed, exhibited or discharged. It is also possible that a classification of leadership can be done in terms of levels based on all these three aspects. We find quite a number of literature devoted to styles of leadership, which really can be avoided as they are not adding anything new. It is quite possible that what we call (can or should call) 'leadership style' can be comprehended in terms of how roles played by leaders are adapted to meet different situations, ranging from sticking to the same roles or focusing on the same functions even when the emerging situations differ remarkably or employ appropriate modifications—sometimes acquiring and deploying new and emerging knowledge and experience—as situations change over time.

It becomes imperative to analyse the three above-mentioned aspects of leadership by way of distinguishing one from the other and demanding that each of these three aspects should be manifest. True, such a manifestation is determined by a number of latent characteristics possessed by the leaders under consideration. The roles and responsibilities of leaders at a given level in a given environment (a complex of people, processes and resources) have to be clearly spelt out by top leadership (or top management) as being compatible with and required for the goals and objectives to be achieved by the organization, feasible against the background of the people who are being picked up as leaders, and enabled in terms of authority provided by the top leadership. Thus, top leadership has the essential responsibility of developing the policy and strategy, desirably involving at least the next lower level of leadership, if not leadership at all levels. In a similar vein, leaders (or managers) at the functional levels will have to develop plans for execution of functions and processes under their charge by involving at least those 'followers' who can be expected to contribute positively. No doubt, to take care of direction and control components, management or leadership at any level has to assign tasks to individuals in the workforce in a manner that is likely to match their knowledge and skill levels to produce the desired results. The manner in which individual workers perform their respective tasks has to be controlled through checks and tests at appropriate stages.

6.4 ROLES AND RESPONSIBILITIES OF LEADERS

We need not bother much about leadership theories in comprehending the importance of leadership in the context of organizational excellence. What are of direct concern are the roles and responsibilities of leaders (at different levels), their interactions with those whom they are meant to lead and the manner in which they deal with problematic situations. While the first depends a lot on the size, the nature and the functions of the organization, the other two are related to traits and habits of the leaders. The second aspect can be segmented into task behaviour and personal behaviour. As can be easily expected, theoretical researches on these aspects are primarily based on real-life experiences augmented and reinforced by thought processes, resulting in diverse, not necessarily disparate, constructs, reflecting different expectations based on different assumptions about human behaviour understood in terms of factor–response relations. The roles and responsibilities of leaders have been brought out in many ways by many experts in leadership. Let us look at the 4E construct, without claiming any superiority of this model over several others. This model is remembered in terms of the four words, each representing one responsibility of leadership: envision, enable, empower and energize. It must be remembered also that in this model leaders and followers are inter-dependent and the responsibilities are meant for both the parties. The first E asks people to envision the future, to generate potent plans that can help the team make things happen, secondly enable the team members through training, coaching or facilitating, thirdly empower the team members to make full application of their capabilities to achieve the expected results, of course within a broad framework and the organizational culture, and finally energize the team members to sustain their motivation, involvement and spirit to drive the entire system.

Going through the huge volume of publications on leadership roles and responsibilities, one ends up figuring out a person with all the virtues and capabilities one is aware of and hence difficult to identify. We try to point out the roles which are crucial to the successful discharge of responsibilities. We should, however, keep in mind the fact that some roles may be just an extension of the corresponding responsibilities, assigned to the manager or leader for onward communication to members of the respective team. For example, a manager/leader has one assignment (may not be clearly expressed) to clearly convey the goals and objectives of the organization and the purpose behind the tasks or activities assigned to the particular team or group or department to fulfil the organization's goals and objectives to all members of the group. Similarly, assigning tasks to individual members and assessing the corresponding outcomes or outputs in relation to the expected ones and taking suitable remedial actions, if needed, to align the outcomes with the desired ones constitute an important responsibility of a leader. Certain leadership roles also involve providing guidance and facilitation to members, as and when needed, which is a crucial role. Quite a few of leadership roles are important to ensure the team's success in achieving the expected outcomes. These roles are quite self-explanatory. Below are listed some commonly mentioned roles of a leader, and a few of them can also be regarded as responsibilities.

Communicator—A leader has to get the goals and objectives of the team under his charge in terms of concrete measures of performance from the appropriate level of leadership and convey the same in clear and simple terms to all his peers. A leader should also remind the team about organizational values as and when needed. He also has to be effective in communications with concerned partners within and outside the organization. The requirement of conveying the desired information without any deviation to the recipient in due time and with due care of the situation in which the latter is placed should be fully met. All this should not give any extra mileage to fluency or any other personal quality of the leader. As a good communicator, the leader acts as a liaison between the workers in the team and the executive management.

Strategist: A leader has to develop a strategy to meet the targeted or expected performance of his team, preferably involving members of the team with adequate skill and experience. Help of any expert from within the organization and even of an external expert, when needed and permitted, may be sought. He has to translate the work entrusted to his (her) team into a concrete and detailed action plan, including whom to assign which task or work element. Right here, he may be guided by an assessment—preferably based on some evidence—of the effectiveness and efficiency of each member who can be considered for execution of a task or activity in terms of speed, accuracy, waste and the like. In fact, an approach like that in solving an assignment problem in optimization may be gainfully adopted for this purpose.

Organizer: A leader has to motivate workers to meet performance targets for the team by setting and meeting their own performance marks. He should be able to unify workers (having their own respective values and ethics) with the organizational culture and the value system of the organization so that they can become effective members of the team. It must be added that this task of unification is easier said than done. It implies to begin with that the leader has conviction in the organization's value system and its ability to meet the organizational goals and objectives. A leader may even negotiate with top executives on certain values and reach mutually agreed modifications.

Facilitator: A leader has an important role to facilitate the work of any worker who seeks help by himself or is found in need of help to carry out tasks assigned to him, by providing necessary guidance or instruction or demonstration. Without emphasizing differences between support and facilitation, workers found unable to meet work standards should be helped by the leader himself or even by someone capable of providing the support needed This role should, of course, be distinguished from the role of a coach where a leader is required or expected to provide some initial training or induction lessons (mostly hands-on) to newly recruited workers or workers assigned new jobs.

Change agent: A true leader should be eager to welcome change by way of innovations and development of new products and services. He also has to be quite flexible and adaptive in regard to new ways of thinking and working.

6.5 LEADERSHIP TRAITS AND ABILITIES

To be effective and to discharge the desired roles and responsibilities, leaders must possess certain abilities and skills, over and above their domain knowledge and relevant skill sets. These have to be acquired by them, cultivated in real-life situations and elevated to become more helpful in enhancing their capabilities to lead. Most of these skills and abilities relate to personal features and characteristics—negative ones to be removed or reduced, positive ones to be upgraded or strengthened. Self-awareness or being aware of one's limitations is the first step. There is no dearth of literature on leadership skills and virtues or abilities.

In a recent blog posted on August 7, 2024, by A. Gomez, 22 leadership skills and virtues have been mentioned. Some of these are really skills, while some others reflect an individual's value system. Also, some of them overlap with some others. The list can be taken as follows:

Communication skill, decision-making skill (ability?), problem-solving skill, accountability (not explicitly mentioned by most experts), motivation, passion, vision self-awareness, empathy, flexibility, learning agility (not generally included), courage, ability to influence (in a positive way), empowerment, delegation (both being really actions or a roles), respect (for others), gratitude (to others), integrity, transparency, creativity, spirit of innovation (by way of support to, involvement in or appreciation of innovation) and empathy.

One comes across a number of publications enumerating such skills, virtues and abilities of individuals who are to be recognized as leaders and required to take the organization forward. As expected, a lot of commonality exists and some distinctive aspects are also appealing.

In this context, we feel that we can concentrate on a set which runs through most of the recommendations. A point to be stressed here is the fact that quite a few experts in management and leadership have unfortunately treated some behavioural traits or habits exhibited by leaders as skills or abilities possessed by them. For example, transparency or integrity is not comparable with communication skill or skill for solving problems. Further, problem-solving and decision-making abilities are somewhat distinct from skills. Transparency is a manifest aspect of behaviour, actions or relations and is rooted in the individual's value system and ethics. Keeping the distinction between inherent skills and abilities, on the one hand, and manifest traits or habits on the other, one can proceed to focus on certain leadership skills on the basis of which leaders are sometimes placed in a hierarchy of levels.

Some of the important aspects of leadership in terms of knowledge and skills acquired, beliefs and tenets followed inherently, abilities revealed in relevant situations and traits displayed in interactions are as follows:

- *Empathy*: ability to place oneself in the position of someone in problem or distress

- *Creativity and innovativeness*: ability to think and act out-of-the-box and to come up with new and novel designs or products or procedures
- *Agility and adaptability*: ability to learn and absorb new ideas, methods and techniques and to accommodate changes in the external environment and in the internal processes due to changes in technology, in human behaviour, and the like
- *Decisiveness*: ability to remain firm in communicating decisions taken after due consideration of all relevant factors and in getting them implemented
- *Self-Awareness*: consciousness about one's limitations in knowledge and skills, and frank admission of the same as and when required (this is sometimes the first priority)
- *Courage* to face undesired eventualities in the right spirit and not to shy away from these.
- *Communication skill* to clearly and convincingly communicate with seniors, peers and juniors, listening carefully and patiently to what they like to say before conveying your views, actions and decisions
- *Team-building abilities* to maintain an atmosphere of mutual trust, respect, help and cooperation, participate in team activities, celebrate team success, etc.
- *Resilience*: ability to recover quickly from the shock due to an unseemly event or its impact, giving priority to the present over the past.

It must be admitted that such a list cannot be exhaustive for all situations in which all different types of organizations may find themselves at some point of time or the other. At the same time, some of the skills and abilities could be specific to organizations and to situations, while most others could be generic. The skills enumerated may have different measures of importance in different contexts.

Next comes the traits or habits of leaders as are manifest in their behaviour, especially at the workplace. A trait or feature of personality manifested repeatedly and consistent on all pertinent occasions becomes a part of one's habit. One is conveniently reminded of Stephen Covey's *Seven Habits of Highly Effective People* suggested in 1989.Of course, leaders are naturally expected to be effective people to begin with. One more habit was introduced in 2004. The author likes to restate the fact that Covey's list is not the 'ultimate', but is definitely quite appealing and to a great extent convincing. Explaining the word 'habit', Covey reminds us that a habit is an intersection of our knowledge, skill and desire. 'What to do' is the knowledge, 'how to do ' is the skill and 'want to do it' is the desire in this triad. Collectively, habits relating to many 'it' (thing that we want to do), on many occasions, justify Covey's remark. "Our character is basically a composite of our habits. Because they are consistent, often unconscious patterns, they constantly, daily express our character." In fact, a great leader is generally looked upon as one with a lofty character.

Coming to the seven traits, we may note that the first three traits relate to the inner self of the leader, the next three address the external world (including the team and other people external to the organization) and the last one is meant both for the leader and the team. These habits find a direct reflection on Goleman's work on emotional intelligence to a large extent. This implies that an effective leader has to be emotionally intelligent. As is known, emotional intelligence (EI) can be enhanced through practices like introspection.

Let us now try to understand the seven traits or habits of leaders, as distinct from leadership styles and roles and responsibilities.

- *Habit 1.* Be pro-active (You are the creator. You are in charge. You should act and not be acted upon.) Focus on what you can control and/or influence and not on what you cannot control or influence. Give attention to your circle of influence where you can take decisions on your own, taking circumstances as constraints and not on your circle of concern where your actions are bound to be guided by the circumstances beyond your control.
- *Habit 2.* Begin with the end in mind. (First creation) Before taking a decision, working out a plan of action or taking an action, keep the desired or the expected result in mind. In some sense, the need for a measure of effectiveness of a decision or an action, as is the practice in operations research, should precede the search for an optimal decision or action. Rather than allowing others and circumstances to shape one's life, this habit links an individual's passion with purpose. Achieving greatness should start from outlining a personal mission statement.
- *Habit 3.* Put first things first. (Second creation) Prioritize and achieve your most important goals instead of constantly reacting to urgencies. A model for identifying and prioritizing tasks should be an essential part in planning one's schedule of activities. Here one can make use of the so-called time matrix with four quadrants in which all tasks before us are classified on two bases: urgency in terms of time and importance in terms of consequence. Quadrant one comprises tasks that are urgent and important like crises and deadline-driven projects, quadrant 2 includes important but not so urgent activities, quadrant 3 corresponds to unimportant tasks that seem to be urgent but are not that important and quadrant 4 is related to unimportant activities that are not to be urgently attended to. Burning out while remaining engaged in quadrant 1 may not be an effective idea and focusing on the next quadrant may be useful. This will ensure the Pareto principle that 20 per cent of your efforts yields 80 per cent of your results.

These three habits which correspond to three distinct components of emotional intelligence—self- awareness, self- regulation and self-motivation—lead to personal victory.

- *Habit 4.* Think win–win. Collaborate more effectively by building high-trust relationships. Win–win philosophy is like a co-operative game and not a competitive one. You have to be empathetic to others; at the same time, you have to be confident that consideration and sensitiveness for others does not detract you as the team leader from achieving the targeted goal. Three vital traits are essential for this paradigm to work such as the following:
  - Integrity is sticking to your true feelings, values and commitments. Maturity is expressing your ideas and feelings with clarity and courage and with consideration for the ideas and feelings of others. Abundance mentality means that there is plenty to satisfy the ideas and expectations of all.

- *Habit 5*. Seek first to understand, then to be understood. (Empathetic communication). Listening carefully should precede making a response by way of an advice or an assessment or a decision. It is not uncommon that a manager is just preparing or readying oneself to spell out one's response while feigning to listen. This is not to happen with a leader. You are required to understand the needs and perspectives of others by carefully and attentively listening to those who have approached you. Otherwise, what you will speak out is expected not to carry any conviction with the others, since what you would say could be irrelevant, unfair or ineffective.
- *Habit 6*. Synergize (Creative collaboration, admitting that the whole is greater than the sum of its parts.). Develop innovative solutions that leverage diversity in ideas and suggestions and satisfy all the key stakeholders. Based on the commitment to search for a solution (to the problem at hand) that we both feel good about, and a deep understanding of each other's points of view, we jointly move to a third alternative that is expected to be more effective than any of the first two. Tools like brainstorming or Delphi exercise are essentially meant to derive the benefits of synergy.
- *Habit 7*. Sharpen the saw. (Renewal) Enhance motivation, energy and seek continuous improvement. Overcoming burnout and achieving work–life balance for renewed activities can be possible through physical, social, emotional and spiritual exercises and can improve capacity for improving. Both physical and mental exhaustion should be avoided and practice of yoga and similar exercises may prove beneficial.

The eighth habit spelt out by Covey in 2004 in his book *The Eighth Habit: From Effectiveness to Greatness* is expectedly a bit more exacting, requiring a lot of reflection and introspection, and emphasizing a strict adherence to principles, to develop self and to help others develop. To quote Covey, it is about finding your voice and inspiring others to find theirs. To find one's voice to be heard by others—for the time being, within the organization—one should develop a mission for one's life, a purpose to be served through an introspection about one's strengths and weaknesses and one's desires, besides a passion to achieve the mission and the purpose and adherence to self-discipline in one's schedule of activities day in and day out. The mission and the purpose should leave room for helping others in finding out who they are and what and how they could satisfy their mission and purpose in life. The core principle behind this eighth habit is commitment to ethical norms and procedures. A great leader has to be a great person with a passionate and disciplined adherence to an ethically supported personal value system aligned with the value system of the organization being served.

In his famous book *Good to Great*, published in 1994, Collins speaks of level 5 leadership as one special feature or trait exhibited by leaders in organizations which have been able to hold on to their positions of strength in the journey towards excellence. Collins mentions five levels of leadership based on traits possessed and practiced. At level 1, one is expected to an effective individual, with requirements for effectiveness which may fall in line with those considered by Covey or, somewhat less ambitiously, in terms of fewer basic traits. At level 2, you become a contributing team member.

To make a cognizable contribution to the team effort, one has to build up relationships with other team members. Going to level 3, the leader is required to organize the team, direct the same towards achieving the set results and at level 4, the individual has to galvanize the team to meet performance goals and objectives. To become a great or level 5 leader, a blend of humility and strong determination is called for. A level 5 leader is always humble, almost shy, willing to admit deficiency in knowledge and skill and even would seek help from team members. At the same time, the leader is highly professionally driven and firmly determined to ensure team success. A difficult balance between humility and determination is demanded for level 5 leadership. Such a leader tends to credit colleagues and/or exterior elements for organizational or team success and accepts responsibility for any failure of the team.

We can browse through the gargantuan volume of material on the subject and try to make a presentation that is by and large complete. However, that may not be mandated by an understanding of the role of leadership as an important driver of excellence. In fact, the nature and extent of knowledge and skill (and even of experience) relevant to the organization and, more so, to its purpose and functions, and the tasks to be fulfilled have not been much discussed as constituting a desired element of leadership, especially at operational levels. The most appealing reason for this can be easily seen as the huge diversity in knowledge and skill set demanded by different organizations and even by different functional areas within the same organization. Cognitive intelligence is naturally taken for granted. However, emotional and social intelligence—concepts and constructs which appeared later—have been linked up with leadership. In fact, emotionally intelligent leaders are the ones who can succeed in this era of merciless, bewildering and disruptive facts and phenomena.

Leaders will have to face different situations, to deal with people with different inherent abilities, individual values and aims and objectives in life, tackle problems faced on different machines and processes with their respective demands for attention. Before embarking on a discussion relating to the roles and responsibilities of a leader, it is quite important to take up some semantic problems associated with the two words 'roles' and 'responsibilities'. Some thought leaders and successful practitioners tend to consider these two words as almost synonymous and even inseparable. There are others who maintain a distinction between these two entities, though some of them agree on the notion of inseparability. The present author likes to point out the distinctive aspects of these two for a better understanding of the situation-specific need of both or one of the two.

Responsibilities are assigned (by people in the higher level in the organizational hierarchy) and discharged, while roles are adopted and played (also displayed) as traits for interaction. Responsibilities are discharged through commensurate authority (often reflected in positions or designations of the assignees), while roles are played on volition by leaders (who may or may not enjoy any authority over peers or subordinates). For a successful discharge of responsibilities, appropriate roles of leadership have to be played. More often than not—and in common situations—managers (at different levels) tasked with respective responsibilities should also play the roles expected of a leader. Of course, it is possible that several teams work under the charge of a manager and each such team has a leader, not identified by the manager but voluntarily accepted by the members of the concerned team.

6.6 EMOTIONAL INTELLIGENCE

Many leadership traits and corresponding situation-specific leadership styles draw upon both cognitive skills and social and emotional skills. Both personal skills and interpersonal skills are called for. In fact, emotional intelligence occupies a more important place in the skill set for an effective leader. It can be argued that a high level of emotional intelligence does imply—may be somewhat indirectly—the trait of humility in level 5 leadership. However, firm determination may not find a direct reflection in emotional intelligence. Emotional intelligence can explain traits of leadership emphasized by quite a few experts. Of the two parallel developments of emotional intelligence—by Bar-on and by Goleman—the latter is more relevant to leadership, and we proceed to look into it.

In the context of an organization, one has to note that an individual is positioned—directly or discreetly, strongly or weakly—among others, as a leader or a peer or a supporting member. In such a situation, intelligence in the individualistic sense as is reported in terms of intelligence quotient (IQ) is inadequate or incapable of explaining variations in 'life success' among individuals across situations.

Right here comes emotional intelligence. In the most general sense, emotional intelligence refers to the ability to 'recognize' and 'regulate' emotions in oneself and in others. This is an innate ability governed primarily by cognitive processes and can be moderated with experience. Goleman and Bar-on argue that emotional intelligence encompasses personality and character, which contribute to life success. Measurement tools in the area of emotional intelligence have taken their shape from preliminary concepts of empathy and alexithymia. Different scales follow different models for emotional intelligence.

EQ has been recognized to predict leadership qualities and to explain the phenomenon that creative individuals like those engaged in development of simple algorithms to solve complex problems or of related software are not always a productive as they could be. Emotional intelligence refers to the capacity for recognizing our own feelings and those of others for motivating ourselves and for managing emotions well in ourselves and in our relationships.

EI (as reflected in EQ (emotional quotient)) describes an ability, a capacity or a skill set to perceive, assess and manage the emotions of one's self, of others and of a group.

EI can be broadly comprehended in terms of five emotional competencies:

1. Ability to identify and name one's own emotional states and to understand the link between emotions, thoughts and actions—awareness
2. Capacity to manage one's emotional states—to control emotions, or to shift undesirable emotional states to more appropriate ones
3. Ability to make good use of opportunities for improving one's competence and to take initiative in handling a situation, self-motivation
4. Empathy or the ability to put oneself in the shoes of others, listening carefully and compassionately to others, sensing the pain by others, accepting views and arguments of others when differences crop up and providing consolation and help to others in need and

5. Social skills to communicate clearly with others, co-operating with others, remaining connected to others, at the same time remaining conscious of one's own strengths and weaknesses.

The first three are personal skills, while the remaining two are inter-personal skills. Five discoveries that leaders have to make in their pursuit of acquiring and / or enhancing EI competencies are

My ideal self—Who do I want to be?
My real self—Who am I? What are my strengths and gaps?
My learning agenda—How can I build on my strengths while reducing the gaps?
Experimenting with and practicing new behaviours, thoughts and feelings to the point of mastery and
Developing supportive and trusting relationships that make change possible.

Goleman argues that for top leaders in our modern and ever-changing business world, EI is four times as important as intellect and expertise. Without EI, the traditional combination of relevant experience and IQ is far more likely to generate a failure than a winner.

The successful transformers of tomorrow will combine IQ-based doer-abilities with EQ-based leadership qualities as they drive and direct the process of change that will make the all-important difference between surviving and thriving for the organizations they lead. They will combine 'management by objectives' with 'management by persuasion'. From a traditional business perspective, empathy may seem inappropriate or unnecessary. However, in today's economy, when team work, cross-cultural sensitivity, coaching and mentoring are essential, empathy is a core skill. 'Alexithymia' motivates us to ruminate over our words [and deeds] used in the past to find out how better we could have performed.

6.7 SERVANT LEADER

The somewhat strange term 'servant leadership' was formally introduced by Greenleaf in his 1970 essay on "The Servant as Leader" and has since been discussed in management literature as a beneficial factor in enhancing the performance of certain types of organizations. However, several educational institutions, religious organizations, charitable trusts and social welfare establishments have rendered commendable services to humanity—quite a few of them in the East—under the leadership and guidance of great personalities who were really 'servant leaders' much earlier than the formal emergence of the concept. Greenleaf felt inspired by reading a novel *Journey to the East* by the German author Herman Hesse and his afterthoughts on the story led him to coin the term 'servant leader'. The basic philosophy is that true leadership emerges from those whose primary motivation is to serve others. The traditional leader thinks of other members as subordinates, 'they serve me' whereas the servant leader feels 'I serve them'.

To elucidate the concept, Greenleaf stated,

> It begins with the natural feeling that one wants to serve first. Then conscious choice brings one to aspire to lead. The best test is: do those served grow as persons, do they, while being served become healthier, wiser, freer, more autonomous, more likely themselves to become servants?

A servant leader puts serving others above all priorities, rather than managing the results, focuses on creating an environment in which the team can thrive and get their highest impact work done. A servant leader, in the words of Spears (1998), can be identified in terms of ten different characteristics such as listening skills as they listen to others receptively to find out clearly their interests, intentions and inclinations, and their perceived problems in workplace and outside. A leader listens intently to others and reflects on the contents whenever alone. They display empathy in putting oneself in the position of others, appreciating their unique spirits and suggestions, and not rejecting those. They practice self-awareness, remaining sharply awake and reasonably disturbed, understanding issues bearing on ethics, power and values healing relationships with whomsoever the leader comes in contact through a subtle communication of the fact that the search for 'wholeness' is common to both the parties persuasion to build a consensus through logical reasoning. The leader uses his personal (not positional) power to convince others and not to coerce them, has the foresight to understand lessons from the past, realities of the present and likely consequences of decisions relating to future. This ability helps the team in making right choices. Leaders are marked by their ability to conceptualize. Conceptualization means the instinct to dream, to come up with ready solutions to day-to-day operational problems and to think about possible solutions to long-term issues, and to strike a delicate balance between conceptual thinking and day-to-day operational approach stewardship to lead the team, keeping in sight organizational objectives and at the same time serving others. The implicit idea is that people cared for contribute their 'best' towards achieving objectives. Leaders also exhibit commitment to growth of people by nurturing personal and professional growth of colleagues. They also adopt Initiatives like making funds available for their personal and professional development, encouraging worker involvement in decision-making, accepting ideas and suggestions from others, helping laid-off employees to find employment are laudable efforts, building community feelings and spirit among members by demonstrating unlimited liability to meet needs and expectations of a specific community within the organization they work 'in' and 'for'.

As mentioned by Spears (1992), servant leadership seeks to involve others in decision-making, is strongly based on ethical and caring behaviour and enhances the growth of workers while improving the caring and quality of organizational life. Several experts, including professionals, considered servant leadership as akin to transformational leadership. While this view is not altogether unacceptable, we have to remember that the focus of transformational leadership is achievement of organizational goals and objectives, while that in the case of servant leadership is growth and development of people (with an implicit impact on organizational performance).

Some critics argue that servant leadership may give out a wrong signal about flexibility drifting towards indiscipline. While remaining friendly with members in the team, the leader has to remain alert and firm to advise and direct any errant member to follow the basic principles defining organizational culture and the spirit embodied in the vision statement of the organization. The very fact that public utility and service organizations are meant to serve others for whom such organizations have been created is sometimes forgotten and even disregarded by the workforce including the leaders who tend to behave as in other for-profit organizations or organizations who act as watchdogs in a bureaucratic fashion.

Servant leadership shows the way for a revamp of organizations engaged in the service sectors, especially those dealing with education, healthcare, support services and the like where several current forms of leadership, propped up by the recent trend in business management, have robbed the organizations of their service orientation culture.

7 Policy and Strategy

7.1 POLICY, THEN STRATEGY

The story of an organization—in the offing, currently growing, functioning under stress or firmly established as one among the best performers in the relevant community—can be comprehended through certain activities carried out by a set of people at various stages. Such activities involve making and implementing decisions by acquiring and deploying necessary resources, including knowledge, skills, technology, and value systems. Decisions and actions in a chain pervade the entire organization.

To begin with, promoters (or architects, or founders or owners) imagine, think of, or dream about some need felt for themselves, for some people they identify or for people at large, and the associated need to set up an organization for meeting those needs. They also think of a broad framework to guide the organization on a continuing basis. All this is reflected in what are termed as the vision and the mission of the organization. A vision is indicative of what the organization would look like and what position it would occupy among its peers in the long run and a mission spells out the purpose the organization would be trying to serve—as desired or intended by the founders or architects. The vision and mission are usually charted out and often written in terms of verbal statements.

It is often argued—and not without any logic—that the long-term vision and the relatively more concrete mission are revealed in a statement of goals and objectives for the organization to achieve. An acceptable distinction between these two, goals and objectives, is not an absolute necessity to move forward, but an attempt towards formulating goals and objectives is be a welcome move to steer clear of any overlaps and redundancies. This explanation is not expected to receive an unqualified acceptance. Goals usually relate to long-term performance, while objectives guide the activities in (relatively) near future. Both can be just verbal statements with no numerical targets involved or can include such targets. Generally, vision statements avoid numerical targets. Goals are usually set for the entire organization, while it is quite common to speak of objectives to be met by segments of the organization, the organization as a whole or for different broad functions. Objectives are usually set in a way that they can be achieved with an efficient use of the available resources, while goals to be achieved may imply augmentation of currently available resources. It should be

DOI: 10.1201/9781003591733-7

remembered that the points of distinction made here are to be considered along with the modifying adverbs like 'usually' or 'generally'.

Following up the vision and mission statements and translating them into a statement of goals and objectives, we need broad guidelines about how do we proceed to achieve these goals and objectives. This results in a policy statement to be carefully crafted, taking into consideration all the relevant externalities like regulatory requirements, societal concerns, situational or locational advantages and hindrances and other concerns. The policy for the organization as a whole sets out certain basic principles to be adhered to and certain priorities to be accorded to different concerns. It is like an instrument to guide the implementation of vision and the mission through the formulation of a concrete strategy. In fact, a policy has often been characterized as a set of priorities and principles, priorities reflected in the mission statement by way of 'thrust areas' within the broad framework of the organization's business processes and principles guiding the organizational culture and value system. Whether the policy relates to the organization as a whole or directly to a function (linked with several others), it should reflect the organizational ethics and values. For example, an organization may discourage or avoid any incorrect or misleading publicity for any of its products or services or even about its internal work culture.

Ethical considerations incorporated in the human resource policy may not allow paying a newly recruited person less than what was committed during the hiring process or may not permit retrenchment before the expiry of the probation period, except when demanded by contingencies.

Policies can be meaningfully classified according to the levels of their adoption and implementation. We have a broad organizational policy and functional/departmental policies. In a manufacturing organization, different policies guide activities in functions like (1) recruitment, retention and retrenchment, (2) procurement through purchase, vendor development and related activities, (3) pricing, distributing and marketing, (4) designing, (5) testing and inspection as parts of quality management, (6) production planning and control and (7) public relations. Within each such second-level policy, there could exist sub-policies to guide secondary and supporting activities.

A strategy is an instrument to put a policy into practice. This is primarily the outcome of quite a few sequential and cascaded decisions to be taken about the core and support processes to be performed by the organization. Usually, the term 'strategy' refers to the level one instrument involving meta-level decisions for the organization as a whole. Instruments meant for second level and supporting policies are often referred to as plans. Referring to the importance of strategy, Toffler (2002) states

> There is a slightly odd notion in business today that things are moving so fast that strategy becomes an obsolete idea. That all you need is to be flexible and adaptable. Or, as the current puts it 'Agile'. This is a mistake. You cannot substitute agility for strategy. If you do not develop a strategy of your own, you become part of someone else's strategy. You, in fact, become reactive to external circumstances. The absence of strategy is fine, if you do not care where you are going.

We may add that a mechanism to adequately and effectively respond to external exigencies as well as to proactively address internally felt need to improve the performance and results may—and possibly should—become an important constituent of a well-worked out strategy.

Policies, strategies and plans involve decisions, and decision-making is quite a comprehensive task very distinct from a decision based on an instant gut feeling. Strategies and plans are quite similar, comparable in terms as they are the instruments for achieving certain pre-specified results and convey how to carry out a function or an activity. But planning has also been an activity by itself, applicable even to policies. In fact, policy planning is a crucial task in the whole process of policy development and analysis.

Like function-specific or otherwise differentiated policies are sometimes spelt out right from the beginning, strategies also can be formulated for specific functions and tasks, besides the overall strategy for the organization as a whole. We also need to develop a policy and implement it through a strategy to manage breakthrough changes in operations and even in the structure. The important point to remember is that policies are intended to provide broad directions, while instruments (incorporated in strategies for respective tasks and at respective levels) are meant for implementation on the ground. Of course, these two are inalienably linked to each other. A strategy has to necessarily take into account the policy and, at the same time, the policy must incorporate issues in implementation.

While it may not be fair to characterize policies and strategies, especially policies reflecting priorities and preferences and certain basic principles of doing business adopted by any organization as a 'static' or applicable for all times to come, some elements of a policy adopted after due deliberations about how the mission of the organization can best be achieved or even a strategy devised for a certain planning horizon should not be changed, except when exigencies so demand. Certain organizational values as well as ethical standards provide organizations with a strength that can help overcome obstacles.

7.2 ROLE OF DECISION-MAKING

The development of a policy or formulation of a strategy (at any level and for any function or task) is essentially an exercise in decision-making—formal and explicit in some cases, informal and implicit in others. Decision-making implies choosing one (or some) out of several identified alternatives for the purpose, which seems to satisfy or meet the purpose most effectively. The choice should also pay due attention to consider only those alternatives which can be implemented, subject to likely constraints on resources available at the decision-maker's end and those imposed by regulatory bodies and external agents.

In fact, policies and strategies have to be planned, and not just spelt out from wishful thinking or after even prolonged deliberations without a proper mechanism for reaching a consensus. Conscious attempts to identify alternatives or options from different sources, including experiences of those involved in the exercise, often produce a sub-optimal choice. The purpose of the exercise in terms of concrete and differentiable aspects of the different alternatives has to be ensured; otherwise the outcome of the exercise (a policy or the corresponding strategy) will fail to be the 'best' or 'best available'. The criteria for comparing the different alternatives have to clearly developed right at the beginning, so that alternatives that are likely to be similar for most of the criteria may be avoided. Planning has been a more generic term referring to an activity that can be carried out at various levels, in various functions, containing varying degrees of detail and resulting in various decisions. Planning definitely involves decision-making. The process of planning can be mapped in terms of the resource inputs (including data), outputs (including decisions to be acted upon), process (of converting inputs into recommended decisions), besides the supplier and the customer. Referring to the literature on operations research or management science, one finds three levels of planning: strategic, operational and tactical. Meta-level strategic planning results in the overall strategy, while operational and tactical plans sometimes correspond to micro-level strategies specific to functions or to tasks.

While policies have been generally developed by the promoters or pioneers through a process of exchanging ideas, thoughts and beliefs—more or less in an unstructured manner—strategies have been formulated more often through formal planning exercises. Obviously, such an exercise has to use a methodology that requires pertinent data and analytics to provide insights into what has been happening and what is likely to happen in the future. A corporate organization-wide strategy is meant to guide the activities of the organization during the planning horizon considered in framing it and the horizon will usually cover a period of five to ten years. Of course, this depends on the nature of the organization, where is it placed currently in terms of its operation, the level of performance achieved thus far and the manner in which the business environment has been changing.

Let us take a look at the decision-making process in its general form, quite applicable to the process in policy planning and strategy formulation. In the classical framework, the decision-makers are presented with a finite set of options or alternative decisions (sometimes referred to as 'strategies') that have been accepted as feasible or implementable and a choice has to be made among these in the face of uncertainty about the future state of nature (business environment in our context). Usually, a finite set of states of nature is envisaged. In decision-making under uncertainty, probabilities for occurrence or realization of different states of nature are known or can be estimated from real-life data, while in decision-making under risk, these probabilities are unknown or cannot be estimated. The outcome of adopting a strategy is affected by the eventual state of nature and hence can only be guessed in advance, though it can be determined only after a state of nature came to be realized. This outcome has been generally referred to as 'pay-off' and thus the decision problem is visualized in terms of a pay-off matrix $((p_{ij}))$, say of order $m \times n$ where p_{ij} is the pay-off to strategy S_i when state N_j is true, $i=1,2,\ldots m$ and $j= 1,2,\ldots .n$ Probabilities π_j associated with the jth state of nature are taken as given in decision-making under uncertainty and as unknown in decision-making under risk.

In decision-making under uncertainty, the optimal choice of a strategy is given by the maximin pay-off criterion due to Laplace to guard against the worst possibility (Laplace is branded as a pessimist), that is, to choose strategy k if the minimum pay-off of this strategy over the states of nature is the maximum among such minima for different strategies. Alternatively, the maximum 'expected' pay-off criterion as suggested by adventurist Hurwicz can be used as the criterion for choice, where the 'expected' pay-off for any strategy is the weighted sum of the maximum and the minimum payoffs to that strategy, with the weight for the maximum taken as α and that for the minimum as $1 - \alpha$. The weight α may be interpreted as the subjective probability that the most favourable state of nature (yielding the maximum pay-off) will come to be true. A third possibility is to convert the pay-off matrix to a regret matrix, where the regret r_{ij} associated with strategy S_i when state N_j has been realized is defined as $r_{ij} = \max p_{ij}$ (over i) – p_{ij}, and to apply the maximin regret criterion for picking up the best strategy. To decide on the optimal strategy under risk, we simply calculate the expected pay-off for each strategy using the probabilities for the n states of nature and choose the strategy with the maximum expected pay-off.

All this seems pretty simple. Problems arise while identifying alternatives (strategies) in terms of distinguishing qualitative and/or quantitative features and checking those for feasibility, imagining different possible states of nature that are mutually exclusive and exhaustive and finally defining and measuring pay-off, also known as the measure of effectiveness. These problems may become hard nuts to crack in the context of organizations operating in a fast and unpredictably changing environment. Honestly speaking, each of these problems throws up challenges to planners and strategists, with complexities depending on the nature of the organization.

In the organizational context, considering the role of decision-making for strategic planning or for developing a strategy either as a corporate organization-wide strategy or as a function-specific management strategy, we have to view the strategy as a set of decisions—sometimes sequential—having an impact on several different aspects of the organization or function related to the goals and objectives to be realized, based on the resources available and the constraints imposed on their deployment. For any strategy to be feasible, alignment with the policy and mission of the organization and implementability within resources that can be committed are essential criteria. A strategy will usually relate to a long time horizon, spanning, say five to ten years, and a whole stream of developments in the economic, political and social situations locally and globally will unfold during the planning horizons. Some of the developments will affect the performance of the organization and the extent to which it will be able to achieve its goals and objectives by the end of this period have to be identified. The path likely to be followed in any such development can be attributed to several factors or events, some internal like change in ownership or top leadership, and some external like a conflict between some countries influencing inter-country trade and movement, extinction of some business partners, especially those in the supply chain, and the emergence of new ones or unpredictable changes in weather conditions. Collectively, these will define the states of nature.

Unlike the simplistic classical paradigm where strategy-state (of nature) interactions are not considered (except by considering some strategies as inadmissible in the context of some particular state(s) of nature), business strategies have to necessarily take into consideration the likely state of nature or the pathway of factors and events likely to influence business performance. This is why strategies are often worked out after the likely future business environments during the planning horizon have been visualized to the extent possible, unlike in the traditional paradigm where strategies are identified independently of states of nature. Furthermore, in the present context, both 'strategy' and 'state of nature' are multi-dimensional and require a vector representation. A strategy would imply a sequence of decisions, one to be adopted as some event takes place or some aspect of the business environment becomes apparent. Each such event may be treated as a node from where many arcs or decision alternatives can emerge and one such decision is chosen on the basis of evaluation of the effectiveness measure (may be in terms of key performance parameters) on each of the arcs. This way, we get the semblance of a decision tree to be visualized as a strategy. Another difficult task is to relate the effectiveness measure or pay-off (a composite of pertinent performance parameters) of the events or the affected aspects of environment in terms of a model. The model parameters have to be necessarily chosen on the basis of experience or intuition.

7.3 SOME DECISION-MAKING MODELS

Business decisions have been and are being taken by leaders and managers using their wisdom, sometimes aided and augmented by some models and methods that may involve quantitative analysis and computations. Some of the well-known models that guide decision-making processes at strategic, operational and tactical levels (including the implementation exercise in some cases) may be looked at in a discussion on policy development and related exercises. As expected, some of the steps are common across all these models, while each model offers some distinctive features and may fit into certain situations. A brief review of a few models follows. As expected, decision-making in business makes use of intelligence, insights and intuition beyond the relevant quantitatively oriented analytical methods and tools. Emphasis on the latter varies and is welcomed by decision-makers to varying extents.

In the Vroom–Yetton model, emphasis is placed on identifying and operationalizing criteria for deciding on the 'best' option or alternative, even prior to generating the options, since some options may not be worth any consideration at all if certain rules are adopted in the beginning to evaluate or judge the feasibility or acceptability of any option. That way, this model differs typically from the traditional frameworks where options or alternatives are first identified (though not necessarily enumerated), some evaluation criteria are developed independently of this task and then the alternatives are evaluated to choose the 'best'.

The Kepner–Tregore approach focusses on possible negative consequences associated with an alternative and on risks, it proceeds in five steps: (1) situation appraisal, (2) problem analysis—problem defined and its root causes determined, (3) decision analysis—alternatives identified and risk associated with each assessed in terms of probabilities of undesired impacts, (4) potential problem analysis—the 'best' alternative is further examined against potential problems and (5) actions proposed to minimize the risk associated with this 'best' alternative. In this approach. emphasis is laid on assessment and prioritization of risks (considering the outcome and the way to achieve it). This may not yield a perfect solution, but we can expect to get the best possible choice. The steps involved in the Kepner–Tregoe model (often referred to as the gold standard in creative thinking skills) can be further elaborated as follows: Establish strategic and mandatory requirements (*musts* indicating what all we must have, operational objectives (*wants* to tell what all we want to have) and restraints (*limits* in the system). In some sense, *musts* and *limits* define what are called 'feasibility' criteria for any alternative.

List as many alternatives or courses of action as may be possible—whether immediately feasible or not. In most business problems, we can reasonably assume a few alternatives, at least initially. This initial set may have to be enlarged only if a desirable solution does not emerge out of this set.

Rate each alternative as Go or Not Go on the basis of each *must* and each *limit*. Rank objectives and assign a priority weight w_j in the range 1–10 to the jth objective. Assign a relative score x_i in the range 1–10 to each feasible alternative to indicate how well the alternative satisfies a given objective. For each alternative i, calculate $\Sigma\ w_i x_i$. Identify the top two or three alternatives based on this sum. List the possible adverse effects associated with each selected alternative and calculate the risk as the product of each anticipated undesired effect and the probability of its eventual realization. Calculate the overall risk associated with an alternative as simply the aggregate of risks corresponding to the possible adverse effects. Calculate the weighted score for each of the top alternatives against the adversity rating and choose the alternative with the least weighted score. In case the probability of realization and seriousness or severity of impact are not available, weights can be taken as 10, 5 and 1 somewhat arbitrarily for high, medium and low risks.

It thus can be seen that Kepner–Tregoe approach bears a lot of similarity with several multi-criteria decision-making methods like TOPSIS. This approach combines decision and action (against anticipated adverse effects). However, the fact that a better comparison among the alternatives could have been done by considering possible containment actions for each alternative has been ignored.

Klein, Calderwood and Clinton-Cirocco (1999) proposed the response primed decision process based on mental simulation of future consequences of a decision and recognition of patterns of happenings in the past. Applicable mostly in emergency service situations like firefighting, this model stresses on experience and intuition, does not attempt an identification of possible alternatives but takes up only one and decides to choose another only when the mental simulation exercise hints at possible risks and adverse impacts. The three steps involved are: (1) experiencing the situation through listening and looking at the situation as it unfolds, (2) analysing the situation in terms of recognizing similar situations experienced earlier and decisions taken there along with the consequences and (3) acting and implementing the decision.

The BADIR framework suggested by Jain and Sharma (2015), duly expanded, stands for Business Questions, Analytics Plan (to choose tools to be used for examining the hypotheses of interest and for extracting useful and intelligent information from data for this purpose), Data Collection (covering relevant data of different types to be collected from different sources through different mechanisms), Insights into what all is happening and how, derived from information extracted from data, to give recipes or alternatives for solving the problem, and Recommendations to be made about the way impacts of the proposed solution on the system under consideration can be realized. This five-step process applies big data to formulate decisions. It starts with the proposition that data exist everywhere, but big data constitute a situation where it is impossible to parse what information is relevant and actionable for our business.

7.4 POLICY PLANNING

Policy planning has been recognized as the primary task in the entire policy development exercise. However, planning has to take care of the fact that a policy once formulated is meant for implementation and further modified, if found necessary through an evaluation process. One point to remember is that usually a policy continues to guide an organization in all its aspects and functions till the policy is revisited on justified grounds. In public service organizations, meant to satisfy some unmet needs for service in the community to be served, the initial policy, in terms of the set of principles and priorities, should be changed if some of the initial needs become satisfied, new service demands arise and new methods and practices to fulfil such growing needs become available. Of course, some basic principles like a focus on customer or beneficiary concerns may endure the life of an organization.
In view of the above considerations, a Policy should be

- Planned within a given framework to meet some specified objectives
- Developed in terms of a sequence of feasible steps
- Accepted by all the stakeholders involved
- Implemented by the concerned group or department
- Evaluated for its outcome or impact compared to those planned for and
- Changed or modified whenever found necessaryolicy formulation combines the first two activities. In this exercise of Policymaking, policy planning occupies the centre-stage.
- Any policy is characterized by its two components viz.
- Intentions spelling out clearly the objectives and the targets to be reached and
- Directions giving out concrete steps for translating the objectives into actions called for. including changes in the existing system.

Intentions flow from the mission statement and directions emanate from principles and priorities adopted. Together, the policy must be aligned with the vision and mission of the organization. The strategy is expected to ensure organizational performance successfully meets the mission requirements. Without clarity in intentions and adequacy of directions, a policy is not expected to deliver the desired results. Targets set without taking due account of constraints on resources—physical, material, technological and human—and their deployment and actions/steps suggested without considering their interdependencies are quite likely to result in the failure of a policy.

Let us focus our attention on an institution for higher education run by a private body, which could be a trust, a society or even a business house. It starts with a bang. The vision statement reads: 'Fostering the spirit of learning and creating knowledge for the benefit of posterity and maintaining one of the best environments for this purpose is our vision for the future.' The mission may be written as 'enabling learners in their pursuit of self-fulfilment and service to society by facilitating their inner abilities to flourish in an enabling environment created and maintained by all associated with this institution who believe in doing things differently.' A cryptic yet hugely meaningful message for an institution of higher learning, considered to be the oldest university of India, reads 'advancement of learning'. Any university or institution for higher education must have been established for a purpose (usually in response to a felt need) with a vision (for the foreseeable future) guided by a mission (setting out the principles and priorities of the organization). A need was felt by some thought leaders more than a century and a half back to encourage, enable and even empower students graduating from the existing schools, colleges and similar institutions to learn new subjects at higher levels and in more effective ways. Arguing that knowledge is the natural output of the learning process, 'creation, conservation and communication of knowledge' may possibly find a place in the vision and mission statement, which somehow has never been formally developed or documented. A fully expanded version takes care of every aspect of higher education as it is to be imparted by the organization. The thrust areas would be clear. Knowledge creation needs research and development work to be pursued by teachers and scholars, conservation mandates publication as well as documentation and preservation, and knowledge should be communicated not merely in the classrooms to the students enrolled, but also to as large a community of willing learners as possible through available communication media and interaction programmes. Here the priority is on what the organization should do and not directly on the outcome to the learners and teachers and to the society at large. Another outcome-oriented mission statement could have been in terms of all-round development—physical, mental and intellectual—of the learners so that the society, the economy and the environment can be effectively and efficiently taken forward by competent and conscientious graduates coming out of the university or otherwise enabled through its outreach programmes.

Let us trace the growth trajectory of an academic institution started by a trust with representatives of several business houses run by some families who had some members who had received more or less good education, acquired some exposure to education systems in developed countries, had taken motivated interest in their family businesses and were of the view that children in families of rich and middle-class and educated people should receive some education that could bring out their inherent abilities in different tasks and could transform them to entrepreneurs in diverse fields of life. Imbued with some conviction and aware of constraints on investment and uncertainty about size of enrolment, the trust started running a non-residential school affiliated to a national registration body, providing facilities catering to the needs of the affluent not available in general schools and fixing a much higher academic fee. The hiring policy was to induct carefully chosen motivated teachers, even successful retirees, who could take up the challenge of providing education in non-traditional ways to foster inquisitiveness among students and enhance their thinking capabilities. Transportation of students from and to their homes was arranged for all students. Teachers were paid well and expected to work full steam for the all-round welfare of the school and its students. The trustees decided on a policy of non-intervention in academic matters directly and left it to an executive officer who would be a link between the school principal and the trustee board. The board had initially acquired a relatively large plot of land slightly outside the main city and slightly inconvenient for access by guardians and other partners. Laboratory and library set-ups were quite up to the mark and discipline among all was remarkable—school results came out to be quite satisfactory and student strength kept on increasing. Facilities for extra-curricular activities and support services were being gradually expanded. True to the adage 'nothing succeeds like success', the trustee board decided to start an engineering college to provide modern education in several branches, including chemical engineering, which was not covered in the curricula of private engineering colleges in nearby places. To foster collaboration with academic and professional bodies, the trust extended academic and logistic support for jointly organizing seminars, workshops and conferences on its campus. While a claim of excellent performance should not be made now by the board, a modest claim of 'good' performance may not be unfair. Then, efforts have to be continued for all-round improvement to realize the transition from 'good' to 'great'.

7.5 STRATEGY FORMULATION

A strategy is a concrete course or sequence of actions chosen (from among alternatives) to satisfy the goals and objectives of the organization, reflecting the principles and priorities embodied in the policy. A strategy has been often regarded as an instrument to enable the organization to achieve its goals as spelt out in the vision and mission statements. Keeping the distinction between the policy and the vision and mission in mind, a strategy takes into account the policy and is aimed at achieving goals and objectives in terms of various processes. A strategy is a concrete and, whenever needed, comprehensive course of action to be adopted along with any restructuring of the organization and a review of its resource base—physical, material, technological and human—followed by recommendations and/or change. A somewhat different view of a strategy has been propounded by Porter (1996). He defines strategy as choosing a unique and valuable position for the organization, rooted in its system of activities that are quite difficult to imitate or to match by others. He focusses on doing different things and on doing things differently. Differentiation lies at the core of Porter's definition. In any case, it stresses on decisions prior to actions that lead to the position of superiority in performance.

A strategy looks into the future—near or remote—and involves decision-making in the face of uncertainty. It has to identify different possible alternatives (courses of action) and examine their feasibility in terms of (requirement of) resources —already available or likely to be augmented—and to assess the outcome of each feasible outcome. This is where problems start. The outcome of any alternative depends on the environment—a set of external factors like market behaviour, regulatory mechanism, co-operation from business partners, overall economic conditions and the like. Changes in any of these factors are quite unpredictable, even if an analysis of their past trends provide some indications. It may be almost impossible to assign probabilities of realization for any of the outcomes and hence in working out the expected outcome which could have been used as a criterion for choice among the alternatives. Taking for granted that the different environment (factor combinations) likely to be realized and that the outcomes for each alternative under the different environments can be found out, theory provides criteria for deciding on the optimal alternative under uncertainty without any probability estimation.

An organization spells out the basic principles for carrying out its business operations and its preferences and priorities in a bid to achieve its mission. A policy could be adopted for each broad function like procurement, production, maintenance, marketing, human resource development, social responsibility initiatives and the like.

The following illustrate situations where policies and strategies have to be carefully distinguished or, better, taken up as a guiding blend: A policy to buy components (for assembly in the organization) versus to make those within the production facility (with due expansion), a policy to maintain a department or group for developing human resources versus to get people trained by established training and development agencies, a policy to market products directly by own people supplying products to distributors or franchisees to take care of sales, having a maintenance department to look after equipment health versus equipping machine operators to take care of routine maintenance and engaging external help for attending to major breakdowns. To carry this point further, let us recall the 'buy or make' decision. Accepting the policy to buy from outside—at least for the contemplated time horizon—several strategic options may exist: (1) identify suppliers who meet your quality, price and delivery requirements as well as potentials for improvements capabilities (and select one vendor or a few, may be using some vendor rating index), (2) select one or few vendors may be with existing limitations but potentials for improvement and start vendor development efforts extending technical and even financial help to meet own requirements on better terms. Similarly, under the policy to be formulated, options would be to acquire a component manufacturing unit and accommodate the same in vertical integration or to expand the existing production facility in terms of equipment, people and other facilities for manufacturing the components needed.

7.6 SCENARIO ANALYSIS

Traditional strategic planning made use of two strong assumptions—proved mostly unwarranted by the recent turbulence in the business environment. These were (1) the future is just an extrapolation of the past and hence future possibilities (outcomes or results from courses of action to be adopted) can be predicted from past results, of course with a certain amount of uncertainty, quantifiable in terms of probability and (2) there exists only one 'optimal' action plan, the role of strategy being to position the organization as close to this 'optimal' as possible. Pitted against this is the belief of those who advocate scenario analysis for formulating a strategy that the future environment will be different from the present one in a manner which cannot be conveniently presented in terms of probability theory. Factors which are certain to change like demographics, technology, basic human wants and needs and global socio-politico-economic situations have to be explored through scenarios. These scenarios are designed to help business leaders cope with seemingly bewildering uncertainty and change, much beyond what they could otherwise visualize. Scenario generation is, by itself, an important task and this task should take due advantage of generative artificial intelligence quite effectively.

Among the many expositions and explanations of scenario analysis, those offered by Ogilvy (2023) and Brander (2023) and a few others seem to be quite concrete and convincing. Ogilvy states—somewhat philosophically—that scenarios provide a way for communities to frame their intentions. They thereby effect 'downward causality' on the present in order to project the present toward a preferred future. Scenarios are narratives or stories about the future told by quite a few members of a team exploring unusual, unexpected and even disturbing pathways affecting business of the organizations, judged in terms of several parameters, including qualitative ones. As usual, a story about the future pathway in terms of happenings on the business horizon some ten years from now will involve many factors and their likely impacts on business of the organization. Evidently, the stories have to hover around the specific aspects of the business of concern to the particular organization. Each scenario describes an imagined sequence of events likely to take place during the planning horizon and the business environment at the end of the horizon. Only those aspects of the environment are to be considered which affect certain features or metrics of performance that have been causing concern to the organization. A scenario, to be completed and put up for further consideration, should provide the means for tracking the trends of factors leading to the events contained in the narrative. It has also to contain enough details to enable assessment of how the scenario once it unfolds will affect achievement of the organization's selected goals and targets.

Usually a whole range of scenarios from the least expected to the most likely, from the most harmful to the most beneficial is developed through freewheeling of ideas, backed by data relating to the past and the present, The team coordinator or mentor has the onerous task of compiling the narratives as they come from the members and putting them before the team for a collective decision on accepting some scenarios for further consideration.

Scenarios generated are screened for their likely or predicted impacts on business performance and only those which concern aspects of business causing concern to top leadership are retained. Then a balancing exercise is taken up to ensure that only scenarios with very positive impacts are not retained and, similarly, only those with high negative impacts are not retained and together not many are retained in the final list. In fact, quite often four to five scenarios including some appearing to be just 'wild' to another and looking quite 'bright' are considered for strategy development.

Once the final list of scenarios has been selected, strategy options for each scenario are assessed and a strategy that can position the organization closest to its goals and objectives is chosen. Finally, that strategy whose components appear against most of the scenarios is chosen. Alternatively, strategy components, which appear against most of the scenarios, are integrated to develop the final strategy. This part of the exercise has been carried out variously by different organizations, some using sophisticated tools and techniques of multi-criteria decision-making to compare different strategy options for a given scenario. Firstly, because a strategy option is a collection of several decision variables whose levels or values have to be worked out and which affect the performance of the organization in some way. Each such variable may justify a distinct set of probabilities or possibilities for its alternate values in order to work out the predicted outcome of choosing a particular value or level. We can, in this way, come a matrix of expected values for each of several performance parameters corresponding to a strategy option. Given, for the sake of simplicity, that we have preferences assigned to the different parameters by way of weights, we can eventually come to overall expected impacts on performance of each strategy option to choose the 'optimal' one. This being a multi-parameter, multi-criteria decision-making problem under uncertainty, we enter into many problems of identifying the parameters of interest, their relative importance measures and the variables which make up for the decision, besides assignment of probabilities to different situations.

While scenarios are generated first and even a handful of them are finally retained for strategy development purpose, followed by development of strategies and choice of a single strategy for final adoption, some experts suggest that possible alternative strategies may be first developed keeping the organization's goals and objectives in mind and ensuring some optimal deployment of resources (physical, material, technological and human) currently available or likely to be available during the planning horizon (spanning, say, five or ten years).The expected impact or outcome on business performance (captured by a few selected parameters or metrics) following each of the strategy options is computed and the best one (using multi-criteria optimization) is eventually selected. Once this is done, scenarios are generated and a strategy selected or even a few potentially good strategies are examined in terms of how many scenarios support a given strategy. The one to be finally adopted should be supported by most of the scenarios. This procedure for strategy development has often been taken as 'stress-testing' of the selected strategy.

Strategy development through scenario analysis differs from traditional strategy selection as equivalent to an exercise in decision-making under uncertainty or risk where strategies have been already identified. Usually a single outcome or effectiveness measure or impact (usually branded as 'pay-off') variable with a known probability distribution is assumed to compute expected values of the variable measure for each strategy under each scenario or environment and a criterion like maximin pay-off or minimax regret or maximum expected pay-off is chosen to select the best strategy. In case of scenario analysis, we consider several pay-offs including some qualitative ones and do not or possibly cannot assume simple probability models appropriate to certain scenarios and the entire exercise is somewhat more involved. Of course, robustness analysis for alternative probability models can, of course, be undertaken.

Whichever long-term strategy is developed, future always uncertain and it may not be wise to hold on to the same strategy as symptoms of change anticipated in the scenarios considered during the strategy development effort. That way, the wise counsel is to remain alert about 'early indicators' of such deviations and to modify relevant constituents of the selected strategy to start with. Guidelines for keeping track of such indicators, aligning them with relevant aspects of the selected strategy, examining the likely impacts of modifications in the adopted strategy, are available in the literature and can be taken advantage of by any organization willing to navigate through disturbances during the planning horizon for a strategy.

7.7 ROBUST STRATEGIES

A strategy involves quite a few choices (among alternatives) or decisions before and after some actions to put the decisions into practice and any choice will result in some output, outcome or impact depending on the environment in which the decision is implemented. The latter is generally uncertain and, hence, it becomes impossible to make a choice and, eventually, a strategy that will yield the 'best' outcome or result, judged in terms of appropriate criteria, is chosen. We have to settle for a strategy that yields satisfactory results in the face of uncertainty. Of course, 'satisfactory' is somewhat relative to the decision-maker. Sometimes, robust decision-making (RDM) combines decision analysis, assumption-based planning, scenario analysis, exploratory modelling to stress testing of strategies over myriad possible paths into the future and then to identify policy-relevant scenarios and robust adaptive strategies. RDM is distinctly different from traditional exercises on decision-making based on expected utility (or expected regret in some situations) using some assumed probabilities. The latter does not involve scenarios, and that too, a whole lot, to do justice to 'deep uncertainty' that RDM attempts to tide over. As has been pointed out by Lempert (2019), analysts use visualization and statistical analysis of the resulting large database of model runs to help decision-makers identify the key features that distinguish those future states in which their plans meet and miss their goals. This way decision-makers can evaluate and modify to choose robust strategies that meet multiple objectives over many scenarios. The traditional expected utility framework characterizes uncertainty with a single joint probability distribution over future states of nature. Such distributions often reflect subjective Bayesian judgement rather than data-based frequentist view. RDM avoids optimality and focusses on robustness or lack of sensitivity as the criterion for choice among strategies. This robustness is not to be narrowly interpreted in terms of some measure of variability in pay-offs or outcomes of a strategy over the possible states of nature usually considered in the traditional decision analysis exercise. RDM is also computation-intensive, but not for predicting the outcomes as efficiently as possible, but to present to the decision-maker as widely revealing a picture about how different alternative strategies can perform in uncertain future states as possible. A robust strategy is one that performs well, compared to the alternatives, over a wide range of possible futures. Such a strategy may not be optimal for a particular future environment which could be the eventual reality.

7.8 IMPLEMENTATION OF STRATEGIES

As pointed out earlier, a strategy (usually at the corporate level) relating to the entire organization (say, to introduce a new process technology or a new software for office administration) or to particular function (like expanding the market in terms of new customer groups) is essentially a plan to achieve certain goals and objectives along with targeted results. The extent to which these results are finally achieved by the strategy adopted will depend upon its implementation. Howsoever skilfully the strategy might have been developed, the end results will be as close to the targets only as effectively as the strategy has been implemented. Actions are involved in 'implementation' and, to be effective, these actions have to preceded by decisions. Put in a different way, these actions have also to be planned. Once a strategy has been developed, additional resource requirements , structural changes or changes in command-and-control mechanisms (in case these are needed), interactions with business partners (if called for) and additional activities to be taken up on a project mode to operationalize the strategy have to be meticulously worked out in detail. Thus, we need what may be called an 'implementation strategy', which is different from 'implementation of a strategy'. Top management has to prioritize these requirements and make sure that the requirements are met at the right time and to the right extent.

An implementation strategy should list a proper sequence of tasks to be undertaken on a project mode, obviously by a project team by involving all the concerned departments or teams. While this list cannot be uniquely spelt out even in a generic fashion, some tasks appear to be quite often relevant and even necessary. We attempt to indicate some of these tasks, which—when properly taken care of—can dispel the notion that implementation of any strategy involves a lot of additional work that may not yield the desired results.

Alignment: The entire implementation project must be aligned with the end results to be achieved through the strategy. Evidently, these results have to clearly understood and communicated among members of the project team as well as among other functions directly or indirectly concerned with the project.

Acquisition and allocation of resources: The requirements of physical, human and financial resources as identified initially have to be acquired in time for deployment by concerned functional management effectively and efficiently.

Accountability: A system of oversight and follow-up, preceded by monitoring, has to be in place to ensure that the implementation project is on the right track and is not unfavourably affected by any misunderstanding of the end results or any lack of cooperation from other teams and functions or lack of coordination among members of the project team. Accountability has to be matched by empowerment of people to the extent they deserve.

Roles and responsibilities: Team members have to be assigned clearly defined roles with no overlap and vagueness. This should take into account specific skills to discharge specific responsibilities, like deploying key performance indicators.

Collaboration: Top management should ensure that the implementation team gets adequate collaboration from those who have to provide support for implementation.

Management support and guidance is essential throughout implementation, from endorsing the implementation strategy to the evaluation of the project in terms of suitable performance measures compared with their desired targets as stated in the strategy to be implemented.

8 Business Process Management

8.1 NEED FOR PROCESS MANAGEMENT

While the roles of leadership and policy and strategy have to be duly acknowledged as the most important drivers of excellence, it must be remembered that excellence is finally achieved only through different business processes are performed or executed and what results they yield. Leadership, through appropriate policy and strategy, is tasked with ensuring high levels of process performance, managing the processes—both core and support— and making efficient use of relevant technology and tools. In fact, process management, involving process analysis, re-engineering and improvement as an aggregate of activities and tasks carried out in different functional areas, is crucial to staying on the track to excellence. This is evident from the growing interest shown by management experts and practicing managers in the subject and books and articles published or accessible otherwise. A remarkable work for the attention of those looking into the future of business process management (BPM) is the article "The Biggest Business Process Management Problems to Solve Before We Die" written by Beerepoot and 41 others published in *Computers in Industry* (April 2023). The article, of course, is highly informative and interesting too.

What is BPM? Is it just the management of the totality of functions and processes executed by the organization through its various segments (departments or branches or groups)? Or is it a process by itself—or, better, an approach—that can and should be applied to all the usual business processes in order to make them more effective in meeting the goals and objectives of the organization and, at the same time, more efficient in making better use of available resources? In fact, it is the latter, as has been clearly articulated by many thought leaders and successful practitioners. As we find in Wikipedia, BPM (with a focus on the word 'management') is the discipline in which people use various methods to discover, model, analyse, measure, improve, optimize and automate business processes.

The BPM of the existing business optimizes workflows through automation and tracking activities, making use of relevant mechanical tools and robots to complete cycles of similar activities, thus enhancing the rate of production, improving quality and reducing manual labour. The automation of materials handling and use of robotics, along with assembly line operations, constitute an important facet of business process management (BPM). Going beyond, BPM does involve process improvement—right

DOI: 10.1201/9781003591733-8

from shop-floor improvements initiated and implemented (with the help of management) by the workers facing operational problems to, in the case of projects, improvements worked out by cross-functional teams in important core or support process or even in several inter-related processes affecting business results. Even bigger changes requiring the existing business model to yield to a new one or changes in the organizational structures, when needed and justified, are all within the ambit of BPM. In fact, leaving aside incremental shop-floor-level, operational improvements which should go on continuously—especially in organizations intending to join the march to excellence—radical changes to bring about improvements have been clubbed under the transformative function termed as 'business process re-engineering' (BPR). The literature on BPR is not scanty and is varied in its coverage focusing on methods and techniques for improvement and on soft aspects like people engagement.

8.2 PROCESS ANALYSIS

Process management should expectedly start with process analysis and typically with a process causing problems either in its operation and/or in its performance (falling short of target). In fact, a proper analysis itself can yield a solution to such a problem, beyond providing vital inputs for the problem-solving exercise. Quite a few simple procedures for such analyses have been tried out in the industry and have been duly documented as well. We consider only two such procedures and just mention a few others.

8.2.1 Process Mapping (SIPOC Model)

The selected business process (not necessarily a manufacturing or even a service process) should be first mapped properly to group its

a. initial and terminal boundaries
b. hard and soft inputs fed into the process as inputs and hard as well as soft outputs coming out of the process
c. suppliers or providers of inputs—internal and external, depending on the process—and customers—internal as well as external—to receive and act upon the output(s)
d. process of transformation of the input into the output and
e. specifications for the inputs and targets for the output(s).

This exercise is usually referred to as process mapping.

Let us consider the case of a machine failing during production and taken for repair (involving replacement of a part or parts) to the repair facility within the production area. For the repair process, the hard input is the failed machine and the soft input includes information about the time the machine failed last, loading on the machine when it failed, nature and the extent of routine care of the machine by the operator. The soft input could also include advice about the time by which the repair process should be completed. The repair process will involve a thorough inspection of the machine to identify the failing components and the failed ones, if any. The repair work is then done using the required materials. The hard output would obviously be the repaired machine and the soft output could be in terms of advice to the user about routine maintenance and load to be placed and a report on the time and cost involved in repair to be forwarded to management.

The process mapping exercise exemplified above is often referred to as the SIPOC model and can be shown in terms of a customer–supplier relationship diagram. SIPOC is the acronym for Supplier, Input, Process, Output, Customer. A detailed process map can reveal value-adding and non-value-adding activities in the process. Such a detailed process map can provide four different perspectives about a process being considered for possible improvement, as indicated below.

Perceived process—What you think the process is
As is process—What the process really is at present.
Should be process—What the process should be to satisfy the customer(s)
Could be process—What the Process could be through appropriate actions.

While the 'should be process' may be difficult to achieve, we have to reach the 'could be process' level. The extent to which the gap between these two entities can be bridged depends on the nature and extent of resources that can be committed to planning, executing and monitoring and correcting the process as is now.

Figure 8.1 shows a process model in which the roles of internal resources committed and external influences affecting the process and its output, and the checks performed on the process and the controls exercised to make the process achieve its targeted output, are clearly indicated.

For the purpose of understanding, controlling, improving or optimizing a process, we need to deal with data on inputs and outputs as well as on the conversion operations. Thus, we need data on input parameters like quality and quantity of input (hard inputs relate to materials, components or products, while soft inputs relate to information regarding the quality of input materials or about the state of an equipment when it failed). Process parameters should include

1. equipment to be used and the relevant parameters and
2. environment under which the process is carried out.

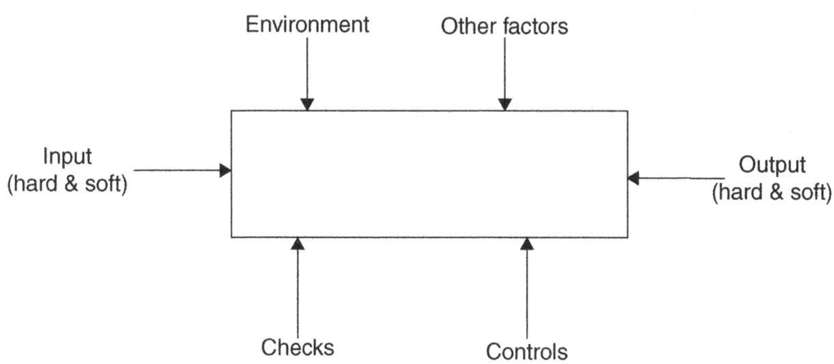

FIGURE 8.1 A process map.

For example, these parameters could be feed-rate of input materials, speed of a machine, alignment between jaws in a cutting tool, time for the operation(s), ambient temperature and humidity. It is well known that the sample size and the acceptance number are the process parameters in a single sampling inspection process. In addition, the features of an item checked during inspection and the method of checking are also important process parameters in inspection.

Output parameters like yield, cost, quality, residual life or availability of the equipment used should also be clearly defined and measured. For the repair process, the time to the next failure of the repaired machine could be an important output parameter. For an inspection process, the number or fraction of defective items left in the lot and found during manufacture and causing problems would be a natural output parameter.

8.2.2 Value Stream Mapping

The 'should be' and even the 'could be' process has to be understood in terms of value addition in each of the steps or activities in a process. Value is what customers need and are agreeable and even able to pay for. A value stream mapping exercise for this purpose reveals how value flows through a process up to the customer and how information flow facilitates the workflow. Activities are classified as value-adding (which have to be performed), non-value-adding (which have to be eliminated or at least reduced without affecting the process) and business value (which have to be retained for business purposes like audit, inspection, etc.). A value stream provides a visual display of all the activities required to produce a product from raw materials to when it reaches the customer, from order to delivery, and a design from concept to product launch. This map pinpoints the steps that should be eliminated or reduced, resulting in a faster and cheaper process.

8.2.3 Deming's P-D-C-A Cycle

One of the earliest and most discussed approaches for improving process quality is contained in the famous Deming Cycle, also referred to as the Deming Wheel. Below we provide a brief outline of this approach:

Plan: Establish performance objectives and standards
Do: Measure actual performance (not just execute)
Check: Determine the gap between actual performance and performance objectives/ standards
Act: Take appropriate action to bridge the gap and make the necessary improvements.

This representation relates to a given process to be executed by an organization. A comprehensive explanation of the terms is provided in the following:

Plan: Establish the objectives and the processes necessary to deliver results in accordance with customer requirements and organizational policies
Do: Implement the processes

Check: Monitor and measure process features against policies, objectives and requirements for the outputs and report the results obtained
Act: Initiate actions to continually improve process performance.

Let us consider the context of a given process, since the spirit of the more comprehensive approach can always be combined with the process-specific approach. The four words—or, better, constructs—in the Deming Wheel (as it is sometimes called) can be interpreted in different ways, possibly depending on the context. Of course, there has to exist a reasonable balance of emphasis on the four constructs.

In the context of process control, the step 'Plan' means specifying how a given process should be performed and how deviations from such a plan can be assessed quantitatively. The process should be controlled at the specified levels or values in respect of inputs, equipment, working environment and also output. However, in the context of process improvement aimed at a greater conformity to the desired specifications or standards, 'Plan' means to work out ways and means to improve operations by determining what is going wrong, causing non-conformity with standards, and then proposing suitable solutions to rectify it. Similarly, the word 'Do' does not simply mean executing the process but really implies that one solves the problems causing deviations from the standards on a small or experimental scale first to see whether the suggested changes in the solution will work or not. It is learning by doing using trial and error to develop an idea about what will eventually lead to an improvement in process performance. 'Check' means to see if the changes achieve the desired results and modifying or refining them, if necessary.

This word was later changed to 'Study'. Both 'check' and 'study' mean putting changes into effect in an organization, thereby pulling people in the direction of making quality improvements. 'Act' means to implement changes on a larger scale, to make them routine and to roll them out to the whole organization, a push activity.

Some authors point out an imbalance among the four constructs. 'Do' and 'Act' seem to be more active, while 'Plan' and 'Check/Study' are more investigative.

At the 'Plan' stage, one has to select the problem, schedule the activities, understand the current situation, set the target, analyse the problem and plan counter-measures. At the 'Act' stage, if the objectives are not met, one has to analyse the results further and plan counter-measures; if the objectives have been met, one has to formalize and standardize the measures adopted. At this stage, the whole exercise has to be reviewed and measures for further improvements in the process are to be planned.

PDSA cycle has been divided into the seven phases of corrective action, in which the phases are dependent on the previous phases and define a cycle. Phase 1 of identifying the opportunity for improvement is followed by the Phase 2 to analyse the selected process and Phase 3 to develop the optimal solutions corresponding to P in Deming Cycle. D in the cycle is Phase 4 of implementing the solutions found in the previous phase. Phase 5 involving a study of the results of the implementation of the improvement actions corresponds to Do in the Deming Cycle. The last element in the Deming Cycle covers two phases: standardize the solution for sustained use and plan for the future.

Six Sigma signifies the outcome of a highly disciplined, top-down, quantitatively oriented, customer-driven, and project-based approach to organizational performance improvement, usually through improvement in strategic business process performance by identifying and solving business problems. Quite often—if not invariably—a business problem is associated with a core or support process failing to completely satisfy the concerned internal or external customer(s), as the case may be. A collection of inter-related processes is also recognized in this context as a process. The problem has been a nagging one and cannot be resolved by a single person in a short time. Hence there arises a need for setting up a project around the problem process to be taken up by a cross-functional team, which has to work for six months or so to work out a desired solution that can fix the problem in the process investigated. As against kaizens, which are meant to solve 'mile-long but inch-deep' problems in the process to work out incremental improvements, Six Sigma methodology is applied to solve a 'mile-deep but inch-long' problem, which when solved, results in a breakthrough improvement that can be sustained.

The route to process improvement in this approach has been encapsulated in the five steps involved in applying this methodology: Define (D), Measure (M), Analyse (A), Improve (I) and Control or Stabilize (C or S). This route emphasizes the importance of the voice of the customer (VoC), measurements of process quality, use of normal distribution and regression relations. The whole idea is that improvement as an activity has to be preceded by definition (of the problem in all relevant details including those for the baseline situation), measurement of different process variables and of yield (fraction of non-defective items or units) and analysis (of yield in relation to process variables). To derive benefits from improvement, the improvement has to be stabilized in terms of holding the process at the levels indicated by the improvement exercise.

DMAIC is essentially an improvement system involving the following steps:

Define—project goals and customer deliverables based on voice of the customer (VoC) and the business problem

Measure—process to evaluate current performance in relation to customer requirements

Analyse—root causes for poor process performance

Improve—process performance and eliminate defects by devising and evaluating multiple solutions. Pilot test the solution and compare performance

Control—improvements, through plans to sustain desired performance.

The following are some important concepts used in DMAIC:

Critical to quality (CTQ)—Attributes most important to the customer (internal or external)

Outside-in approach—Looking at internal processes from the customer's perspective.

8.3 BUSINESS PROCESS RE-ENGINEERING

Floated as the prime mover of breakthrough improvement in organizational performance, especially in underperforming ones, by Hammer & Champy (1993), business process re-engineering (BPR) came to occupy the centre stage in management practice since the close of the last century. However, BPR failed to live up to its expectation in many organizations, which had invested their resources in this new-found initiative. The reasons included a wrong BPR implementation strategy, misalignment of the IT architecture, inflexible management team, workforce mired in outmoded technologies and work practices, lack of adequate training for executives and workers in BPR methodology, and finally, lack of a long-term commitment on the part of top leadership.

The steps involved in BPR will appear to be—as it should—very similar to those used in any continuous process improvement effort, which is covered in subsequent sections. The following is quite illustrative of how a BPR strategy can be developed and implemented.

Identify the process to be re-engineered. It should be critical in terms of its performance level affecting the overall organizational or at least operational, performance and one which has to be made more effective in achieving its targeted performance, more efficient through a better utilization of resources and streamlined to ensure ease of operation and almost effortless repetition of operation cycles without a hitch.

Analyse the process, starting with a process map and examining the workflow as well as mapping the value stream to identify operations to be improved, corrected and eliminated. A comprehensive data-based analysis of the process under consideration as a critical component of the organization's business process is basic to BPR.

Redesign the process to track and remove the bottlenecks, inefficiencies and redundancies. To this end, a creative approach involving all the stakeholders, from the workers directly engaged in the process to the internal customers receiving the process output, from executive management concerned with the process to top leadership worried about the process performance, is required.

Test the new process as designed above to ascertain if the initially identified problems have been successfully resolved or if problems still persist. In the latter case, a second round of troubleshooting exercise through critical thinking has to be taken up to eventually come up with a process design that can achieve the desired level of performance.

Implement the newly designed process by documenting the changes initiated and by training the workforce in the new process operations. The documented re-designed process should be communicated to all concerned.

The re-engineered process should be continuously monitored for its smooth operation, inviting no new or unforeseen difficulties, and delivering the performance target in repeated cycles.

Related processes and even other processes to be rectified should also be subjected to re-engineering in a desired sequence so that organizational performance as a whole can be improved.

Re-engineering with a new design should run concurrently with exercises at the operational level to identify problems of gradual slowdown in performance due to factors not visualized initially or cropping up only after implementation and affecting the performance of the process. BPR can take due advantage of IT and IT-enabled services in process analysis, redesign and stabilization aspects. In fact, BPR can be quite logically seen as a step in technological innovation, as discussed in Chapter 10.

BPR makes use of simple tools generally. Softwares exist for most of them. Tools like process mapping or workflow analysis as well as value-stream mapping have become highly standardized. Similar is the case with benchmarking. Automation tools using robotics have to be process-specific. For process modelling and determining optimum levels of process parameters, efficient software-based algorithms can provide necessary help through simulation and optimization exercises.

8.4 AI-ENHANCED BPM

The applications of Artificial Intelligence (in its initial stage) and of AI-assisted systems offer immense help in business process management. In fact, many of the steps involved in BPR can be facilitated largely and even strengthened courses of action, decided by decision-makers, augmented, improved and implemented quickly, correctly and efficiently by using simple AI-enabled processes, including automation. In this context, AI can be taken simply as the simulation of human cognitive functions using intelligent agents. We can accept the definition of AI systems offered by Organisation of Economic Co-operation and Development (OECD) in 2019, which stated, "AI systems are machine-based systems that can, for a set of human-defined objectives, make predictions, recommendations or decisions influencing real or virtual environments. AI systems are designed to operate with varying levels of autonomy". AI systems have since become more intelligent, capable of taking over roles of human decision-makers and, surely, of operatives in different spheres of an organization. In the domain of service-provider organizations, particularly those engaged in education and health care, – AI and AI-assisted systems have been playing crucial roles in improving the core processes to a large extent.

AI in BPM refers to the application of AI technologies and tools in business processes—support processes as well as core processes—to enhance efficiency, accuracy and timeliness and to assist and augment and even improve human decision-making. Towards this, cognitive automation (learning from relevant data and improving a process iteratively), machine learning and deep learning, computer vision and pattern recognition, natural language processing and other AI-powered tools can be deployed along with robotics and internet of things (IoT) for streamlining operations, reducing manual effort, avoiding human errors and making more informed decisions for optimizing performance in different business processes.

Professionals in the field have pointed out the beneficial roles of AI and AI-enabled systems in several distinct areas of process management, such as in re-engineering a process, like process simulation, process mining, process automation, process optimization and process monitoring and continuous improvement. An element of process discovery through capture and analysis of different aspects of process operation and related results in process performance can reveal hidden efficiencies of a process, which may be exploited for redesigning the process.

Integration of AI-enabled systems has resulted in enhanced performance in support processes like human resource management including talent management needed to institutionalize innovation, finance capable of generating real-time figures of merit reflecting performance in people-related, society-related and environment-related areas, marketing focused on customer behaviour. In fact, intelligent decision-making

making use of more than usual streams of data, analysis of structured data using tools and using AI tools to gather decision options, in addition to their predicted outcomes and recommendations for decisions to be taken by humans, affects business processes and organizational performance to a large extent and therein lies the role of AI.

8.5 NEED FOR CONTINUOUS IMPROVEMENT

Leadership and resources along with due focus on strategy, customers, workforce and effective as well as efficient deployment of resources are extremely important in a journey towards excellence. The key to excellent performance lies in continuous improvement in all business processes—both core and support. Just an improvement—which could be major breakthrough achieved at the end of a protracted and challenging exercise—can achieve 'success' and result in an expected improvement in performance, particularly in the results. However, to sustain such an improvement in the context of growing knowledge related to processes, plans and procedures, the 'improved' state resulting from the breakthrough effort has to continue in terms of smaller exercises resulting in incremental improvements. Such marginal improvements are absolutely necessary to sustain the outcome of a major improvement and, hence, in the journey towards excellence.

Improvement in all aspects of the organization—particularly in all its processes—has to be ensured through the participation of its entire workforce irrespective of the levels and the processes on which they work. Those who are directly involved in executing the processes and the activities that are undertaken as part of the processes are the best persons to identify problems with the current procedures and suggest the best possible solutions as well. These solutions may need verification and validation by executives or leaders and implementation of any suggested solution confirmed to be effective in resolving a problem has to be the responsibility of management. In the entire exercise, workers have to be encouraged to come up with problems and suggested solutions and even empowered in certain cases to implement the solutions they suggest. This has the additional benefit of employee engagement. Any effective solution suggested to resolve a problem should be acknowledged and rewarded. All this can work as a continuous exercise towards problem resolution resulting in improved performance.

Quite a few of the methods and tools for continuous improvement owe their origin to quality improvement efforts by Japanese industries and subsequently modified and adopted by manufacturing as well as service organizations in many countries. They are basically simple, easy to operate and found effective in most cases. These methods obviously bear Japanese names. Several such methods and tools have been modified from time to time, discussed and documented very widely and have helped many organizations to grow and excel on a sustained basis.

Continuous improvement can be pursued as an objective and achieved through both breakthrough improvements at the end of a substantial effort by usually a group–often multi-disciplinary—engaged in removing a crucial bottleneck in performance and incremental improvements on the shop floor as the result of small group activities using simple tools and coming up with solutions to smaller obstacles to work or in the work environment. Methods like Six Sigma with the DMAIC or DMADV steps or an approach like lean management are illustrations of the first type. Use of Deming's P-D-C-A cycle, kaizen and its recent extensions, design of experiments and study of response surface to determine the optimal process design, just-in-time approach for inventory management, poka-yoke or mistake-proofing and quite a few others have been found pretty useful to work out moderate or marginal improvements.

8.6 SUSTAINING IMPROVEMENTS

Improvements, especially organization-wide radical or breakthrough improvements, have to be sustained through function-specific or shop-floor level improvements through modifications and adjustments that are carried out almost routinely. Without these seemingly unimportant incremental improvements, benefits from breakthrough or major improvements cannot be derived. Some of the simple tools for this purpose, as mentioned above, are discussed briefly in the following, starting with kaizen, which has gone through several versions, and those that can bring about moderate improvements.

The basic principles of kaizen (*kai* means change and *zen* means good) were formulated in different ways by different exponents making use of the same fundamentals. Stated as ten principles, kaizen involves the following:

Never stop.
Be proactive.
Eliminate old practices.
Don't stop.
Make corrections on the go.
Empower all employees to speak up.
Don't assume new methods will work.
Practice the 'five why' method.
Crowdsource to learn more and hear more.
Be economical.

The spirit of kaizen or "doing little things better everyday—slow, gradual but constant improvement—continuous improvement in any area that will eliminate waste and improve customer satisfaction" is nicely brought out in the following poem:

For the want of a nail, the shoe was lost
For the want of a shoe, the horse was lost
For the want of a horse, the general was lost
For the want of a general, the army was lost
For the want of an army, the battle was lost
For the want (loss?) of a battle, the war was lost
For the want (loss?) of a war, the country was lost
And all for the sake of a nail.

The scope of kaizen has been expanded in recent time to kaizen blitz, which is an expedited formal and structured improvement by a team working on the lines of DMAIC but limited usually to a process, leading to a situational improvement.

The five S principles for developing a productive work environment involve five steps: sort (*seiko* or *seiri*) to segregate unnecessary items and eliminate them, place each item in its place (*seiton*), shine (*seiso*) by cleaning the workplace and the equipment, standardize to ensure that the previous steps can be carried out in the same manner following standards which are developed by involving the shop-floor people (*seiketsu*) and sustain benefits of the above four activities through discipline, commitment and empowerment (*shitsuke*). In recent times, these principles have been extended to include safety and others to result in 7 S. Some western organizations have branded these principles as 5C's: Clear, configure, clean and check, conform and customize for sustained practice. Some even use the acronym CANDO, implying clean up, arrange, neatness, discipline and ongoing improvement.

Just-in-time arrivals of incoming materials or work-in-progress items from the previous workstation to the next process ensures a steady flow of work and minimizes inventory of finished goods or work-in-progress items. It is an inventory management principle which implies close interactions with suppliers so that incoming materials arrive exactly when they are needed. The principle (JIT) is also extended to the dispatch of finished goods to customers.

Single-minute exchange of dye (SMED), pioneered by Shingo, is a tool that reduces the time to change over production equipment. SMED steps take place in one of the two areas: internal, which happens when the equipment or process is stopped, and external when the same is running. The steps involved in SMED are Identify the process, identify elements of the process, separate the external elements, convert internal to external and streamline.

Zero quality control (Judoka), initiated by Toyota in the early 1900s, and focusing on built-in quality, works on the transfer of human intelligence to automated machinery so that machines are able to stop, start, load and unload automatically, detect when a defective part has been produced, stop themselves and signal for help. This implies operators are freed up to do value-adding work.

1. It operates on four principles: Detect an abnormality, stop the process, fix the immediate problem, and investigate and remove the root cause.

Production work cells: The work done at each stage of production is expanded so that a team of workers is responsible for a stage of production and has the authority to make minor changes to procedures within the confines of a time limit and standards. (The autonomy enjoyed by operators in Toyota was in direct contrast to Ford's production line drones. Empowering the workers to take corrective actions within certain limits meant that there was less need for inspectors to stop mistakes from occurring.)

Kanban (meaning 'sign board') is a scheduling system introduced by Ohno (1998) to limit the build-up of excess inventory (work in process). Kanban cards ensure that parts are made only in response to demand. Each workstation must have a card before it can start any operation. Problem areas are indicated by measuring the lead time and cycle time for the full process and the component process steps. Beyond the production area, kanban is also used in the transportation area to authorize the full container to the downstream workstation.

Mistake-proofing (poka-yoke), introduced by Shingo (1989), is a procedure that prevents defects or malfunction during manufacture by, for example, eliminating choices that lead to incorrect action, stop a process if an error is made and prevent machine damage. It is essentially a safeguard that prevents a process from proceeding to the next step unless proper conditions have been satisfied. Three different methods—contact method, fixed value method and motion-stop method—are employed in poka-yoke, each acting as a warning or a control device.

Several simple tools are used to work out meaningful improvements in processes or in elements thereof. These are also used in methods to bring about breakthrough improvements at the scale of an organization. These include fishbone diagram, root-cause analysis, matrix data analysis, listening to voice of the customer, quality function deployment, design and analysis of controlled experiments and response surface analysis.

8.7 LEAN MANAGEMENT

Lean is a management philosophy manifest in a work organization permeating the entire organization. With a direct reference to 'no frills' or unnecessary work elements, lean management does not have a necessary linkage with the shrinkage of workforce. A redundancy in work and hence in resources deployed may call for a reduction in size, not otherwise. Rooted in the Toyota Production System and enriched by quite a few industry majors in the United States, and formalized by researchers at the Massachusetts Institute of Technology, the word 'lean' was coined by John Krafcik in his 1988 article "The Triumph of Lean Production System" and the lean philosophy was introduced formally by Womack and Jones (1996) in their famous book *The Machine that Changed the World.* The core of lean management and related movements characterized by terms like lean manufacturing, lean enterprise, or lean Six Sigma and similar others is to minimize wastes of different kinds and build values in products and services to satisfy felt and even anticipated needs of customers. It all started with lean manufacturing and quite rapidly pervaded other areas of activity, not merely within manufacturing and conventional service enterprises, but also to completely different activity areas, most importantly in the area of software development.

The basic objective of lean management is to create and deliver value to the customer by optimizing resource utilization and to maintain a steady workflow based on real (and not just anticipated) customer demand. It fosters shared responsibility, encourages multi-skilling of workforce, respects individual contributions to quality and ensures continuous improvement through a gradual elimination of all types of wastes—both resources and activities—thereby leading to enhanced efficiency and profitability. Thus, an organization with an urge to achieve excellence can ill afford to ignore the basic principles of lean management, even if it does not put in place a system called lean management.

Lean management can be introduced in an organization in a 'project mode', but once incorporated as a work management culture, it goes on to effect continuous improvement in all activities related to procurement, production and delivery. It is distinct from Six Sigma projects, which are distinct and occasional activity sets focused on a particular process or a group of inter-related processes that has been identified as causing business problems. Outsourcing of certain jobs or tasks may be financially or otherwise beneficial to some organizations, which may claim to have become 'lean', but this strategy has not been promoted as a corollary of 'lean management'. This implication is particularly important in situations where the outsourced jobs or tasks have the potential to influence customer satisfaction. Balancing financial results against customer concerns is indeed a tightrope walk for management in some organizations.

A few experts in production management have attempted to portray lean manufacturing and lean management itself as a form of scrum management wherein the entire production process is broken up into segments, which are taken care of separately in terms of access to and analysis of all relevant information about the segment, followed by an inspection, to find out any deviations in results realized from those to be achieved and adaptation of changes or corrections to remove these deviations. The focus is not to start with a strategic overall plan for the entire process and to review and modify, if needed, at the end but to work in terms of plans for each of the segments separately and revisit plans for different segments on the basis of tollgate reviews.

This involves breaking the total work process into a collection of smaller parts, visualizing the workflow and removing process interruptions and roadblocks. Rework and even replanning are not ruled out in scrum management, while the emphasis in lean management is on doing right the first time in each work element and minimize, if not eliminate, any rework.

The basic principles or steps involved in lean management can be stated as follows. Over time and through meaningful and customized adaptations, these have undergone a few changes, though the fundamentals remain the same:

1. Identify value (in products or services), which leads to customer satisfaction and even customer delight for which customers are willing to pay (provided they can afford). As is known in value analysis, value associated with a product or service can be categorized into three types: use or functional value, prestige or esteem, and exchange or re-use value. The relative importance of each type depends on the product and, to some extent, on the customer. While use or functional value is most often the most important factor affecting customer satisfaction, household decorative items demand more of esteem and prestige value. Certain products like a watch may call for high levels of both these two types of value. A piece of household or office furniture or—far away from this—a software developed for a modest purpose may attach a lot of importance to exchange value or re-use value. With a given product line and a comprehensive interaction with existing as well as potential customers, the organization should filter the voice of the customer to identify the types of value appreciated or likely to be appreciated by customers. The factors underlying these aspects of value addition have to be sorted out and the product or service design should be worked out by using a method like quality function deployment.
2. Map the value stream to find out how and to what extent value is being added at each of the workstations through which the product flows. This exercise

will help in identifying work elements which do not add any value in the absolute sense and can be easily avoided without causing any dissatisfaction of the customer such as elements which do not add any value but cannot be conveniently avoided now and elements which add value that customers do not insist upon and to which customers are indifferent, and finally work elements that do add value which affect customer satisfaction to the extent that removal of such a work element will result in customer dissatisfaction. The objective should be to completely eliminate the first category of work elements which are really wasteful and to try gradually to do away with the second category.

3. Establish a steady workflow that is not hampered by any production-line bottlenecks in terms of either delayed arrival of in-process work items from the previous station or delayed outflow of work-in-progress items from the present workstation to the next in line. The first one leads to a loss of useful machine time in waiting for inputs, while the second one causes the same problem to the machine next in order.
4. Establish a 'pull' factor to deploy resources only when needed to avoid overproduction, sometimes at the cost of running the production equipment with a higher than usual load or faster than prescribed. which may lead to avoidable storage of work-in-progress items or finished goods. A tendency to produce in anticipation of a prospective order is justified only on the basis of a due analysis of past experiences and to a permissible extent to meet shortages when demand arises in bulk and has to be met in a short time. Finished goods in inventory not against any order may not fetch any revenue or may fetch a discounted revenue because of a delay in realizing the payment.
5. Pursue a culture of perfection in which every worker in each workstation does execute the work element with no defect or deviation from the plan, so that no rework is involved, not to speak of rejects. Both rework and rejects imply resources, which could have been avoided if the work was done right the first time.

Wastes of different forms have to be carefully revealed, since besides being often overlooked as being trifling, these could be hidden from view and have to be searched for. The general tendency to ignore minor wastes apparently of no significance in the context of value addition has to be countered by the appreciation of lean philosophy, which is not merely providing value addition at each workstation but also doing this efficiently by avoiding wastes of resources (materials, machine/equipment time, man-hours, physical facility like space, etc.). The elimination of wastes is the primary goal of any lean system. In the words of Domingo (2007), lean declares war on wastes. Waste or *muda* is anything that does not have any value or does not add any value and is something the customer will not pay for.
The seven forms of *muda* are the following:

Waste of overproduction (largest waste)
Waste of time on hand (waiting)
Waste of transportation

Waste of processing itself
Waste of stock in hand
Waste of unnecessary or avoidable movement
Waste in making defective product items

Muda exists along with two other undesirable features to be avoided: *mura* and *muri*. *Muri* translates as overload or overburden on resource input. It includes the undesired impact of workers working beyond their physical capacity or a production equipment used beyond its rated capacity. Such uses can lead to a sudden breakdown or a gradual decline in output in terms of quantity or quality, or even both. The results could be extremely undesirable in some cases. *Mura* refers to unevenness or the absence of a steady workflow at a constant rate till the time the rate is changed to meet the requirements.

If different elements in the process or different workstations do not produce outputs at rates to maintain a balance between the output of one being just the input for the next at the right time, there will be production-line bottlenecks. As such, *mura* and *muri* may not directly be reckoned as wastes, but they may lead to wastes in several ways.

Some other wastes came to be recognized over time and one of them is nonused employee talent. Taken in a broad sense, the fact that there are employees with specialized knowledge and/or skills that are not being currently utilized in terms of the current deployment pattern should be considered as the eighth form of *muda* or waste and should be avoided. This point was discreetly hinted by Deming while he spelt out the 14 deadly diseases that the Japanese industry has to avoid. Knowledge mapping and knowledge-based deployment should be taken up seriously by the human resource development function within any organization trying to achieve excellence.

Eight tools and approaches were used in the Toyota Production System to prevent wastes and losses and some of them have been touched upon in the previous section. Total productive maintenance is taken up in a separate section.

8.8 BREAKTHROUGH IMPROVEMENT

In recent times, quality management—and, for that matter, management in general—places greater emphasis on comprehensive problem-solving methodologies rather than stand-alone tools and techniques that are useful for analysing and improving a few aspects of a 'business problem'. Thus, methodologies like Six Sigma, 8-D and SAMIE are widely employed. Of these methodologies, Six Sigma can be regarded the 'mother' and the others are like its derivatives. We briefly describe here the Six Sigma approach and indicate how 8D and SAMIE can be conveniently related to it. A detailed discussion on Six Sigma approach appears in the next section.

To solve nagging business problems, especially those associated with internal business processes, Six Sigma approach provides a structured methodology that originated in a few American industry organizations and has since been pursued in many organizations through several customizations and modifications to become a pretty well-known methodology for process improvement, even in service enterprises. This methodology has been characterized consistently, on top of modifications and applications, in terms of five steps indicated by the acronym DMAIC: define, measure, analyse, improve and control (sometimes replaced by stabilize, S). In the area of design development (for the purpose of new product development), the steps get modified as define, measure, analyse, develop and verify.

It has been aptly pointed out that as against kaizens which are meant to solve 'mile-long' but inch-deep problems in the process to work out incremental improvements, Six Sigma methodology is applied to solve a mile-deep but inch-long problem, which when solved, results in a breakthrough improvement that can be sustained. It may be noted that this methodology is primarily aimed at reducing variations or enhancing consistency in quality across units (of product or service) so that a vast majority of the units produced meet relevant requirements and are defect-free.

The 8D methodology is not much different from the Six Sigma methodology for solving a business problem and achieving a significant improvement in operational performance. It involves eight steps—referred to as eight disciplines: Set up a team, (1) define and describe the problem, (2) develop interim containment plan, (3) implement and verify the interim action, (4) determine, identify and verify root causes and escape points, (5) choose and verify permanent correction for problem/non-conformity, (6) implement and validate corrective actions, (7) take preventive measures and (8) celebrate and congratulate the team.

- In fact, the current 8D methodology is a variant of TOPS (team-oriented problem-solving) that was introduced in Ford during 1986–87 and later incorporated in Mil-Std 1520 titled Corrective Action and Disposition System for Non-Conforming Material.
- In 1990s, Ford came out with a revised version, which is the current Global 8D.
- There is a lot of similarity, as expected, between Six Sigma and 8D methodologies. But there exist important differences as well.
- While 8D is a team-oriented approach for solving even recurring problems, Six Sigma is a top-down project-based team approach to solve business problems that are usually first encountered.
- The biggest difference is in Step 3 of the 8D process: implement and verify interim containment, with no comparable step in the DMAIC process.
- We should note that interim containment may give us a false sense that the problem is solved and that we can move on to the next problem.
- Interim containment is typically a quick and faster band-aid approach to addressing the symptoms of a problem and not the root causes.
- Of course, reality is such that if a business problem is serious enough, it merits a band-aid to 'stop bleeding' right now. Then it is up to the team to complete the rest of the 8D process, removing the band-aid, once the root cause is found and addressed.

The SAMIE approach (Chang, 1993) involves five phases: select, analyse, measure, improve and evaluate. Some may argue that the second and third phases should be interchanged, since analysis will involve measurements. The model can be adapted to suit organizational requirements. Thus the first phase may not be required when a particular process is known to be the first candidate for any possible improvement. And the model can be adopted even for a process that has been currently productive satisfactorily but leaves room for improvement. The five phases are:

Select: 1. Define key requirements for core customers.

2. Identify the process to be improved on a priority basis.

Analyse: 1. Document the process as it is now.

2. Establish process performance measures.

Measure: 1. Gather baseline process performance data.

2. Identify and quantify process performance gaps.

Improve: 1. Set process improvement goals.

2. Develop and implement improvement measures on a trial run basis

Evaluate: 1. Assess the impacts of process improvements.

2. Standardize process and monitor ongoing process improvements.

8.9 SIX SIGMA METHODOLOGY

A widely practiced approach to process improvement in both manufacturing and service organizations including R&D establishments and academic institutions, Six Sigma initiative has been aptly characterized as a top-down, highly disciplined, process-focused, quantitatively oriented and project-based approach for improving the performance of a business process that has an important role in influencing organizational performance. Unlike methods and tools used by small groups of workers to solve work-related shop-floor-level problems, this approach is adopted by top leadership to solve a business problem arising from a critical process or a set of inter-related processes. Of course, even to solve problems with an underperforming process, other linked-in processes also have to be covered by the problem-solving exercise. Top leadership identifies the problem and sets up a multi-disciplinary multi-level team to work on a project mode to find out a solution, try it out and work out necessary modifications for eventual adoption. The approach uses probabilistic and statistical tools as are necessary along with relevant technological tools. The approach is highly disciplined where the sequence of steps is scrupulously followed, with no step missed or taken out of sequence. The steps involved in this approach are duly captured by the mnemonic DMAIC (or S) or DMADV, D to indicate Define, M implies Measure, A stands for Analyse, I denotes Improve and C means Control (S to mean Stabilize).

DMADV refers to a design development exercise where the last two letters stand respectively for Develop and Validate.

An improvement project identified by top management as influencing business performance significantly is assigned to a cross-functional team. The problem is associated with a process or a set of inter-related processes. The problem is not a trivial one and demands a comprehensive and dedicated effort of a multi-disciplinary team to solve. In fact, such an exercise is taken up as a project. The project works necessarily in terms of the five steps spelt out in the acronym in the prescribed sequence and through the accepted processes of each step. The methodology takes due advantage of quantitative methods for analysis of process data (existing or generated through properly designed experiments) and improvement of the process(es) and/or the operations involved are suggested based on the analysis. The first phase or step starts with identifying the process(es) causing the business problem and mapping the same in terms of inputs and suppliers, process operations, outputs and customers (internal or external). Listening to the voice of the customer using a suitable documented mechanism is critical. The scope and purpose of the project should be clearly spelt out, preferably in terms of numerical targets against which project performance can be assessed. The scope emerges from the customers' end and from the process operations, and the targets could relate to increased customer satisfaction, reduced cycle time, reduced cost of non-conformance and of wastes and, eventually, to enhanced profit. Also, the time by which the project should be completed and the cost that should not be exceeded have to be specified.

The second phase provides a comprehensive picture about the baseline situation in terms of the nature and frequency of occurrence for each type of problem or adverse consequence. It has a strong focus on 'defects', which are deviations from the process plan or relevant requirements) that renders the unit under consideration unfit for use (by the end-user or the next-in-line operator). In order to avoid wrong decisions to classify a unit as defect-free when it has indeed one or more defects or to classify a defect-free unit as defective, the measurement system (applied to the underlying quality characteristic), a measurement system analysis in terms of gauge R and r (Reproducibility and repeatability) has to be taken up. The process capability for each process and sub-process (whenever relevant) has to be worked out by using a standard measure like Cp or Cpk on a sample of units for each of the quality characteristics under study.

Once data have been compiled, these have to be analysed to find out cause–effect relations by way of regressions or root-cause analysis. Whenever needed, tests of suitably framed hypotheses have to be carried out to establish the relative importance of each factor affecting any chosen response variable aligned to any of the targets spelt out in the first phase. Several statistical methods and software facilitating their applications to the data at hand are used in this phase.

The fourth phase is to develop several alternatives to improve the current performance level as judged by some of the targeted objectives of the project. These are mostly in the form of process setting, relating to process parameters as well as to equipment parameters, besides ambient conditions. Requirements on inputs may also be relevant.

For this purpose, properly designed experiments suitable for finding the optimum factor-level combinations in a relatively small number of repetitions or replications have to be carried out and variations in response variable(s) have to be analysed. Response surfaces have to be explored and using appropriate methods, the optimal or better-than-others treatment combinations have to be worked out to be recommended for process improvement. Allowing for the usual process slippages, robust combinations using evolutionary operations method introduced by Box, the final recommendations can be made.

The last phase of trial runs are carried out to verify if the expected improvement is really worth implementing, bearing in mind the effort and associated cost of resetting the process and equipment parameters and imposing restrictions on inputs. This task is sometimes absorbed in the previous phase. However, this verification is essential for adopting any improvement suggested. The accepted changes have to be properly documented, standardized for convenient practice and monitored during implementation. Whenever needed, suitable modifications have to be effected.

Being project-based, Six Sigma methodology involves formal and structured reviews at pre-identified stages called tollgates, which guide the project team by assessing the progress on completion of a stage against pre-specified objectives and targets and either signals a go-ahead to the next stage or proposes an exercise to remedy the shortcomings during the current stage. It might also involve going back to a previous stage or shifting the corrective action to a subsequent stage. A review meeting is to be preceded by the preparation of relevant documents, which is circulated among members. Different situations justify different tollgates during each of the five phases of the methodology. One interesting set of three tollgates in each of the five phases is recommended by Villanova University and will be briefly stated here. During the Define phase, these relate to approval of the project charter, finalizing needs and requirements of customers (both internal and external), and mapping of the process(es). Identifying indicators of process performance, planning the data collection procedure and preparing a document on the baseline situation are the three tollgates during the second phase. The third phase involves reviews at the end of a Root-cause analysis, consisting of a narrow root cause analysis to pinpoint what contributes most to the problem under consideration and critical root cause analysis focused on the few causes that have to be eliminated or controlled. However, these three can be broadly put under root-cause analysis; the second tollgate should be placed after other ways of analysing the data and the third should come at the end of interpreting the outcomes of analysis. The three tollgates identified during the Improve phase correspond to generating possible solutions to the problem to be resolved, choosing one of the possible solutions based on interests of the customer, the organization and the project, while the third and the most important tollgate is related to the implementation of the chosen solution. During the last phase, the first tollgate corresponds to developing a plan to monitor and control the proposed solution (essentially a change in process settings). Potential problems that may arise after responses are generated are reviewed at the second gate, while the third gate during this final phase relates to the plan for sustaining the improvement through necessary modifications from time to time. As remarked earlier, this entire delineation of toll gates is just indicative and not conclusive, to be finalized for each case separately.

To appreciate the quantitative orientation of this methodology, we should remember that the focus here is on drastic reduction, if not complete elimination, of defects arising in different operations within the selected process(es). As is well known, such defects as are detected during process inspection lead to avoidable wastes of materials, machine time and man-hours involved in rework and repair besides rejects. Such defects need control over process settings, primarily machine/equipment settings and input material. Relevant process/equipment variables affecting quality characteristics under inspection and material variables have to be identified and nominal standards and permissible specifications have to be noted. Right from the identification of defects to defining the response variable(s) for optimization based on data generated through experiments, which have to be effectively designed to the subsequent study of the response surface, followed by an assessment of improved situation, suitable statistical tools are adequately used in the Six Sigma approach. What follows is very brief and partial account of these tools.

Let us consider a process involving k operations. At each stage, a certain number of units U are produced; for each unit, a number of properties are checked to identify defects D. O is treated as the number of opportunities for a defect that crops up in a unit inspected. The numbers will vary from operation to operation and from one unit to another. These are summarized in terms of

Defects per unit DPU = D/U
Defects per unit opportunity DPO -DPU / O
Defects per million opportunity DPMO = DPO × 10^6
Throughput yield (fraction of defect-free units) Yp = exp (DPU)
Rolled throughput yield (for the process as a whole) = (YRT)

This corresponds to a value of the standard normal deviate or *z*-score. Sometimes, organizations these days calculate a ppm defect rate or DPMO and then convert this to a Six Sigma measurement unit that considers a 1.5σ shift in the process mean to account for usual process drifting. Sigma quality level with this shift can be approximated as

Sigma quality level = 0.8406 + [(29.37 – 2.221 × In (ppm)]%

This means if we control our process to ensure that units produced by the process lie within mean ± 3SD and the proportion of defect-free units is 0.9973, we are working at a 3.43 sigma level. The desired sigma quality level is 4.50.

Once we note that the rolled throughput yield has to be maximized through correspondingly optimal process setting in respect of the process variables, we have to design and execute a screening experiment to find out the factors or variables which affect the yield (defect-free units).

A Six Sigma organization is not just a 'tool-pusher'. At the same time, Six Sigma implies a highly disciplined and quantitatively oriented approach to performance improvement. Quite naturally, several mathematical and statistical tools and techniques are involved in this approach. However, the nature of process analysis and optimization determines the types of such tools and techniques to be used in a particular project. Some projects may need only simple tools for data collection, presentation and interpretation. Some other projects may call for the application of sophisticated and even not-so-well-known tools and techniques.

Since most of the variables—categorical or continuous—in process analysis and subsequent optimization are affected by a multiplicity of causes and are consequently unpredictable, they are random and hence the need to apply probability calculus and statistical methods for dealing with such variables. Broadly speaking, we need the following types of analysis and hence the corresponding tools and techniques:

Dependence analysis is performed through categorical data analysis, correlation and regression studies. Depending on the situation, we may have to use non-linear regression models or logistic regression models. In fact, screening of factors to be eventually included in the experiment can be based on tests of significance applied to the regression coefficients.

Analysis of factor means response relations in terms of ANOVA (analysis of variance) and ANCOVA (analysis of covariance) along with their multivariate generalizations. Subsequently, we have to express the yield (observed response or its transform thereof) in terms of a regression with those factor variables which significantly affect yield. This provides us with the response surface, which has to be explored to find out the optimal design point as closely as possible.

Optimality analysis is generally taken up in a Six Sigma project to enable methods to reach the optimum point on the response surface or to find the best combination of levels of controllable factors, besides undertaking quite a few other supportive analyses and a few techniques for data collection and presentation, including exploratory data analysis.

Since the Six Sigma approach talks of the defect rate m (called DPU or the number of defects per unit, noting that a defect corresponds to a non-conformity that begets customer dissatisfaction) and yield Y, that is, the proportion of defect-free units, we make use of the Poisson probability model to derive yield in terms of the defect rate as $Y = \exp(-m)$. Of course, the normal probability model is used extensively in carrying out most of the analyses mentioned earlier. Sometimes, data have to be collected from a continuous stream of incoming materials or in-process items or finished goods and a sampling rate has to be worked out. This rate or fraction may even be allowed to depend on the results obtained sequentially.

Techniques like ANOVA (analysis of variance) are pretty useful in various contexts like establishing measurement system capability by calculating repeatability and reproducibility of measurements or test results. Speaking of designed experiments where ANOVA and ANCOVA (analysis of covariance) have to be used to analyse the resulting data, we need to define the experimental material and the experimental units and the response variable (s) appropriate to the context. Somewhat sophisticated designs are needed in some cases to minimize effort or maximize efficiency in the estimation of effects. Quite often, we have to use fractional factorial experiments to minimize the number of design points. For optimum-seeking experiments, sequential experiments may be preferred. In some cases, mixture designs are the appropriate ones.

To apply methods for locating the optimal combination of factor levels or the optimal design point from a study of the response surface, we need to know and use the method of steepest ascent or other forms of gradient search method. The problem becomes quite complicated if we deal with more than one response variable.

Tools used in the context of hypothesis testing and estimation based on parametric models or on a distribution-free approach have to be applied to evaluate and compare effects, to establish relations and to yield ideas about improvements. Dealing with attribute or categorical data, we need to adopt slightly different procedures. And since normality is assumed in most estimation and testing procedures, it will be wise to test for this normality itself before proceeding further.

Incidentally, for this problem as well as for many others, there are no unique tests or estimation procedures and one has to make a wise choice of any procedure in a given context. Presently, many of these inference procedures take account of any prior information or knowledge about the underlying parameter(s) to make the procedures more effective. The desirability of opting for such sophisticated procedure involving many choice problems like that of the prior distribution (s) has to be carefully examined, since there would always remain the need for an early decision that is at least admissible.

8.10 TOTAL PRODUCTIVE MAINTENANCE

Consistent with the 'lean' philosophy and almost a corollary of lean management is the concept of Total Productive Maintenance (TPM) introduced by Nakajima in 1971 and subsequently translated into a wholesome practice through the Japanese Institute of Plant Maintenance. This now stands for an organization-wide maintenance system covering the entire life of the equipment range for production. The word 'total' in TPM implies

1. Total or, rather, overall effectiveness (considering productivity, cost, quality, delivery, safety, environment and motivation) of equipment has to be achieved.
2. Total participation of all employees from plant-floor operators to executives involved in planning, purchasing, installing and caring for equipment is ensured and
3. Total maintenance system including preventive maintenance, breakdown maintenance and autonomous maintenance has to be in place. The objective is to eliminate all avoidable wastes, which can be attributed to inadequate or imperfect performance of production equipment.

The objective behind TPM is to avoid six big losses:

1. Breakdown losses
2. Set-up and adjustment losses
3. Losses due to idling and minor stoppages of equipment
4. Loss due to operation at reduced speed or at less than full (design) load, that is, reduced capacity loss
5. Loss due to poor product quality (rework time)
6. Loss due to start-up/restart to the point of stable production

Equipment performance is judged by a metric called overall equipment effectiveness (OEE) defined as

OEE = Availability × Performance Efficiency × Quality Rate

where

Availability = Actual loading time / Available loading time, the Actual loading time being the available time minus the down time.

Performance Efficiency = (Theoretical cycle time × process amount) divided by the operating time and

Quality Rate = (processed amount – defect amount) / processed amount.

Some problems may arise in defining the available loading time since loading may not take place on scheduled holidays even if the equipment is available. Similarly, if a planned maintenance activity is taken up, its duration should be taken out of the available loading time. The point is not to 'penalize' the equipment for some problem on the part of the user to load the equipment. Performance efficiency takes care of speed of processing and requires a rational setting of the theoretical cycle time for processing a certain amount, especially when the amount fed is rather too large or too small.

To be a concrete and actionable concept, OEE needs a break-up of the total time on a machine in terms of the following relations:

Total Time = Total up-time (during which the equipment is available for loading or is in the functional 'state') also called the total available time + total downtime (during which it is in the failed state and is undergoing repair or is waiting for repair)

Total Available Time (AvT) = Total time during which it remains loaded (LT) + total time for planned rest due to preventive maintenance or production holidays etc.

Total loading time (LT) = Total operating time during which it is processing the job (OT) + breakdown time during operation (BT)

Operating time (OT) = Net operational time during which it works with scheduled speed and scheduled capacity + time during which there is speed or load loss

Net operational time (NOT) = Value-adding operational time (VAOT) + time for rework for defects to be removed.

In terms of the above break-up, the factors defining OEE can be stated as

Availability (Av) =OT / LT = (Loading time – Downtime) / Loading time

Performance rate (PR) = NOT / OT = (Ideal cycle time per unit x Number of units produced) / Operational time

Quality rate (QR) = VAOT / NOT= (Actual process yield (number of units produced)) /

(Number of units to have been produced or design yield)

Finally, OEE= Av × PR × QR.

It thus becomes clear that planning of schedules for rest periods, preventive maintenance and loading times and for speed and feed are important considerations and that all these are data-based, thus calling for adequate attention to collecting, recording and analysing the relevant data. In parallel, autonomous maintenance entrusting operators with the task of routine maintenance by way of cleaning, lubricating and fixing accessories on providing them with necessary material and knowledge help stands out as a primary activity to ward off breakdown during operation. An 85 per cent OEE is taken as world-class and a benchmark to be established for a typically high manufacturing capability.

Nakajima (1988) points out five pillars of TPM as follows:

1. Losses resulting from imperfect equipment performance have to be targeted for removal. These are (a) breakdown due to equipment failure, (b) long setup and adjustment activities, (c) idling and minor stoppages, (d) reduced speed in processing, (e) defects in processing and (f) reduced yield between machine start-up and production.
2. Planned preventive maintenance involving schedules inspection and service to restore equipment vigour and effectiveness taken up during a work shutdown period.
3. Autonomous maintenance or maintenance performed routinely by operators to prevent deterioration. This requires operating equipment correctly, maintaining basic equipment constitution, making necessary adjustments in jigs and fixtures, accurately recording data on breakdown and malfunction and collaborating with maintenance personnel.
4. Training and education as a natural corollary of autonomous maintenance. Operators to acquire basic engineering knowledge and skills to operate equipment properly and to imbibe basic skills to carry out routine maintenance, maintenance staff to be trained adequately in advanced maintenance practices involving shock pulse monitoring, vibration analysis, lubricant oil debris analysis, etc. Both operators and maintenance staff as well as other employees to receive training in soft skills to realize the need for achieving high levels of performance on all machines to satisfy both internal and external customers.
5. Equipment improvement and maintenance prevention activities initiated by the management in the first case and by all operating and maintenance staff in the second are essential to reap the benefits of a high overall equipment effectiveness. Decisions to purchase new equipment and to install equipment in the right locations and emphasizing on regular check-ups on a routine basis followed by cleaning and proper upkeep of equipment play important roles.

9 Innovation and New Product Development

9.1 INTRODUCTION

Continuing from the need to pursue a policy of continuous improvement in all functions influencing the performance of different activities and in the consequent results, the importance of 'innovation' to come up with new products and services is being felt increasingly in a world marked by stiff competition and a host of uncertainties. In fact, without innovation, staying ahead of the pack may not be possible currently and not so even in the near future. At the same time, we should keep in mind that innovation and invention are different entities, that unlike inventions which are credited to single gifted individuals or to small groups of such individuals, innovative spirit can be cultivated as an organizational value. Indeed, innovation has to be institutionalized and not kept as the preserve of a select few creative minds within the organization.

Innovation is by itself not synonymous with excellence, rather it is an essential trigger to put an organization on the right track to excellence. Innovation results in a new, novel and non-trivial product or service, which is market-ready, and hence can benefit the organization financially. It leads to a win–win situation, the unmet needs of the customer are met or, at least, greater utility is derived from the product produced from an innovation exercise and the producer was able to expand the markets or derive justifiably higher profits. Innovation usually relates to new product or service development, though it may also imply a non-trivial and even significant improvement in an existing product or service in terms of better quality, longer life and easier and safer use and maintenance.

Innovation necessarily involves imagination, creative thinking, strong common sense and application of relevant technologies. It is more often a team exercise, rather than the brainchild of a highly creative and well-versed member working in isolation.

Economic benefits from innovation and new product development have been pointed out by many experts and realized by many organizations across the spectrum. One such realization is reproduced below.

"Wealth in the new regime flows directly from Innovation, not from Optimization—wealth is not gained by perfecting the known, but by imperfectly seizing the unknown", says Kenneth Kelly in *New Rules for the New Economy*. These two entities, however, are not mutually exclusive. They often go hand in hand. This statement has to be

DOI: 10.1201/9781003591733-9

accepted considering the words 'innovation' and 'optimization', which can, in some sense, differentiate between these two important contributors to improvement, particularly in the context of technical systems, in various directions and dimensions. In fact, one of these two words (concepts) may imply the other in many situations.

A 'technical system', which can correspond to a concrete product/object or a service, is meant to carry out 'intended functions', which imply some benefits, in an existing or anticipated 'environment', using resources, involving some costs and causing some undesired effects or harms.

- Given the intended functions, optimization may aim at cost minimization and therefore resource usage minimization and reliability maximization.
- Given the resources, innovation may aim at enhancing the functions, even offering new products and services.
- Innovation may focus on functions not currently delivered, but felt as needed or even imagined/dreamt and may draw upon resources not hitherto tapped or appreciated.

While innovation is sometimes triggered by imagination or fantasy, is referred to in stray incidents and anecdotes in mythologies and fairy tales, is generally associated with or attributed to creative minds and are concrete outcomes of abstract innovative ideas, optimization is far more down- to-earth, relating only to existing systems and usually follows well-trodden paths.

While optimization attempts to reach 'perfection' in the sense of a maximum or a minimum using a criterion as the basis, 'innovation' believes in progressive realization of 'ideality'. Optimization with respect to a criterion or objective function leaves no room for further optimization.

'Innovation' in one go is only an imperfect realization of an imagined 'ideality'. In fact, technical systems evolve through the 'innovation chain'. Improvement means 'doing something better' and coming up with a more effective product, process or service. Innovation means 'doing something different', not necessarily related to the existing concrete entities. Improvement can also be brought about by innovation and innovation can effect further improvement in a novel way.

9.2 INNOVATIONS DEFINED AND ILLUSTRATED

9.2.1 What Is Innovation?

As delineated above, innovation can be defined in many ways, at least seemingly and partially. Suggested from different perspectives, several such definitions accessible on the web and quoted below collectively represent a refreshing picture.

- "Innovation is Fresh thinking that creates value New products, business processes and organic changes that create wealth or social welfare" ("Something New Under the Sun: A Special Report on Innovation," *The Economist*, October 13, 2007, p. 4).
- A change that creates a new dimension of performance.

- A creation (a new device or process) resulting from study and experimentation; the creation of something in the mind; the act of starting something for the first time (both suggested by Peter Drucker).
- Any creative idea, getting implemented or realized successfully (M. Soltynski)
- An imaginative activity fashioned so as to produce outcomes that are both original and of commercial value.
- The practical translation of ideas into new or improved products, services, processes, systems or social interactions. (Think Smart Blog)
- A venture away from familiar ground into uncharted territory. (Yost)
- The net result of thinking differently, thinking deliberately (in specific ways) about existing problems and unmet needs. (A.W. Ulwick (2005). *What Customers Want*)
- The embodiment, combination and/or synthesis of knowledge in novel, relevant, valued new products, processes or services. (A. Razeghi, (2008) *The Riddle*).

In short, an innovation is something original, new or improved (in some sense) that creates value and breaks into (or obtains a foothold in) a market or society.

Innovation is an imperative for sustainability—be it for a manufacturing /service organization or for the national economy as a whole. In fact, some say that there are only two functions of business: Innovation and Marketing. Invention is directly related to the creation of new knowledge—ideas and methods. Innovation makes use of novel ideas and methods to offer a concrete entity. Inventions may imply new abstract principles or 'laws' and contribute to basic sciences. Of course, the steam engine was also invented. Inventive ideas by themselves are not innovations and this perception implies Innovation = Invention + Enterprise.

9.2.2 Innovation and Improvement

Improvement means 'doing something better' and coming up with a more effective product or process or service. Innovation means 'doing something different', not necessarily related to existing concrete entities. Improvement can also be brought about by innovation and innovation can effect further improvement in a novel way.

9.2.3 Innovation Illustrated

Innovations can be found in products, processes and services that we use. Some of them may appear as simplistic, as if not worth being called 'innovations'. Most electronic and telecommunication gadgets we use today were great innovations. Simple innovations that are taken for granted today include a pencil with the eraser at the bottom, a hammer with a nail-puller, a child's stroller also a shopping cart, a toothbrush also a tongue-cleaner or a mobile phone also a camera. There are too many of them.

Innovators are not always great scholars or established professionals. They think over mundane problems and make thoughtful use of known solutions to problems in a related field. Such innovations may not be widely known. A few such simple yet engaging innovations are mentioned below without naming the innovator:

- A bed for sick, bedridden people, fitted with a remote controller, that can move up and down and swing left and right.
- A wheelchair that can climb stairs. There are five small basic wheels, one of which could just climb one stair. On the half-metre long bracket in front, there are another four wheels. The chair has 14 wheels, can be stopped at any position, can move forward and backward and rotate at one position, can climb 2,992 stairs at a time and weighs just 54 kg.
- A wristwatch for ladies, wired to give a shock to anyone attempting to molest her, giving her a chance to flee.
- Children invent, engineers fabricate, companies commercialize—an inverted model for innovation. Innovation contributes to social good and to national wealth, benefits many by enhancing their access, capacity to afford. The use of goods and services improve the quality of life by offering new goods and services and, thereby, creating new jobs to meet the demand of an expanding market.
- Towards this objective, the spirit of innovation has to be promoted among many in the workforce and innovation has to be institutionalized, so that it no longer resides in the solitary recluse of a few highly imaginative and creative individuals with substantial thinking and knowledge capabilities. This will enable an enterprise to design, develop and deliver new products and services to improve quality of life.

9.3 NEW PRODUCT DEVELOPMENT

Whatever is produced by an organization, including a service, using an existing and known design or a newly created design, is a product. A new product may imply a product

- New to the market
- New to the company
- Completely new to create new markets.

Industry leaders look upon NPD as a pro-active process, in contrast to a reactive strategy in which nothing is done until problems occur or some competitor introduces an innovation.

The NPD process involves the following:

- Idea generation—ideas for new products may be obtained from basic research using a strength, weakness, opportunities, threats (SWOT) analysis, competitors, focus groups, employees, market and consumer trends, salespeople, corporate spies and trade shows. Idea generation involves brainstorming.

- Idea screening: The second step is to eliminate unsound concepts prior to investment of resources on them, raising questions like
  - -Will the customers in the target market benefit from the product?
  - What is the size and growth forecasts of the market segment/ target market?
  - What is the current or expected competitive pressure for the product idea?
  - Is it technically feasible to manufacture the product?
  - Will the product be profitable when manufactured and delivered to the customer at the target price?

- Concept development and testing: To develop marketing and engineering details, the costs of manufacturing and reactions of a few potential customers to the idea are tested.
- Business analysis: The following are estimated: likely selling price, sales volume and break-even point for quantity/volume to be produced in the first go. This fifth step has to be carefully planned and carried out using data, experience and intuition.
- Beta testing of a physical prototype or mock-up (and its packaging) in typical usage condition constitutes the next stage. Market testing should follow in terms of focus group customer interviews and trade shows. Necessary adjustments are made, an initial run of the product is made and the product sold in a test market.
- Technical implementation becomes the next step, involving resource estimation with special reference to resource augmentation, logistics plan, supplier collaboration, engineering operations planning. This is a comprehensive multi-party exercise, requiring interaction with external players.
- Commercialization involving product launch, promotional efforts and filling the distribution pipeline is expectedly the final step that may necessitate revisiting some of the previous steps to yield the expected benefits from NPD.

NPD as a process is closely linked to quality improvement, which is not always explicitly visible. This requires a broad view of product quality revealed through performance during its use by the customer. If NPD targets a completely new product, quality improvement does not become an apparent corollary. In other cases, NPD is linked to quality improvement.

A very well-accepted approach to NPD and/or quality improvement is provided by TRIZ (in English TIPS, the theory of inventive problem-solving) developed by G. Altshuller (a Russian scientist-cum-engineer) and tried out successfully by several leading manufacturing and service organizations across the world during the last two decades. Originally concerned with 'technical systems', TRIZ applications now cover a wide range of services.

In most cases of NPD, we are raising the ideality of the underlying 'technical system'. Generally, this can be achieved by resolving some physical or technical contradiction. In fact, when we talk of using new materials and new ways of processing such materials in developing a new product, we really have a conflict resolution done, though not consciously. A brief discussion of TRIZ follows in the section.

9.4 THEORY OF INVENTIVE PROBLEM-SOLVING (TRIZ)

While it is very true that innovative ideas translated through an appropriate enterprise result in new products (services also being recognized as products), and innovative ideas cannot be taught or put forward in a similar manner, the possibility of suggestions to seek such ideas and to put them on trial can be facilitated by using some relevant concepts, methods and tools even software. A theory combines all these and one such theory was propounded by G. Altshuller, a Russian scientist-cum-engineer. Altshuller strongly condemned the idea of re-inventing the wheel, advocated the use of inventive principles extracted from a deep analysis of numerous 'patented' inventions and firmly believed that reference to a data-base on standard solutions to generic problems can help solving specific problems through 'systematic innovation thinking.' TRIZ, along with the algorithmic theory of design and optimization methods, do lead to new and improved designs for 'technical systems'. TRIZ has gone through many modifications and have been successfully applied in many corporate firms. As its very name implies, TRIZ is a theory complete with concepts and definitions, methods, tools and techniques. A comprehensive understanding of TRIZ and allied topics justifies an assimilation time of at least a week.

We all appreciate that wealth (in a broad sense) flows directly from innovation, that innovation heavily draws upon creativity. Creativity is possessed to varying extents by all of us, and creativity and or the innovation mindset should be duly nurtured. Young and talented individuals should be exposed to concepts, methods and tools as well as databases pertaining to guided innovation.

The theory of inventive problem-solving (TRIZ or TIPS), proposed by the Russian engineer-cum-scientist Altshuller, provides insights into innovation and some basic concepts and tools which can be imbibed by smart students who have a zeal for innovation. TRIZ speaks of a 39×39 contradiction matrix, a set of 40 inventive principles and a set of 76 standard solutions to complex problems, resulting in innovations. Some of the 40 inventive principles are really interesting and can guide talented individuals to become innovators.

9.4.1 Concept of Ideality

In the context of innovation, assumed to take place in stages to justify the concept of an innovation chain, as discussed in TRIZ, the concept of ideality provides the guiding principle for a step-by-step innovation exercise or a problem-solving exercise. In fact, a measure of ideality of the concrete entity (a 'technical system' which is mostly a product but can also be a service) is expected to increase as innovation proceeds in steps or stages. In fact, this enhancement in ideality, understood broadly to imply an improvement in any complex device or equipment that has greatly impacted human life, is somewhat evident. The addition to or augmentation in the functionality of automobiles, aircrafts, elevators, mobile phones, digital cameras, surgical implants, imaging and scanning devices and a whole host of other products along with the associated additional conveniences and safety features amply demonstrates the principle that technical systems evolve towards greater ideality.

While ideality should and does increase as we try our hands at successive efforts to 'innovate' or to improve upon an existing product, we should also think about innovation resulting in a new product—of course with a modest view of 'new'—and about ideality right in the first stage in the innovation chain. In other words, we have to appreciate that the outcome of an innovation exercise reflected in some measure of 'ideality' is one step towards satisfying the innovator and convincing the interest groups involved in the innovation exercise.

Ideality, like creativity, does not admit of an unequivocal definition, especially in terms of a blanket definition that embraces all types of 'technical systems' in manufacturing and service domains. We discuss some of the issues in defining and, more so, in measuring ideality in concrete terms based on evidence that are or can be made available.

Before attempting to define ideality, let us note that an ideal system does not exist, but all of its functions are fulfilled at the right time and at the right place—without energy, substance, or other resources, and without any ill effects.

Ideality, at any point in the chain, has been generally taken as the ratio between the [aggregate of all functions (that can be carried out by the entity as developed in the current stage) with positive or beneficial impacts] and the [aggregate of costs incurred or the life cycle cost along with the harmful impacts on the use environment and the general environment].

Petrov and Seredinski (2005) define the idealization level as

$I = \alpha \Sigma F_i Q_i / [\Sigma C_j + \Sigma\beta_k H_k]$ where F= useful function (effect)
where Q= quality of useful function
C=time and mean cost for useful function implementation
H–nuisances and α and β are 'accommodation coefficients'.

Evidently, this definition or some variant of this definition, like one that incorporates consequences of public goods or free resources, raises the following questions:.

- How to identify and, subsequently, to quantify the different elements like 'functions' and 'impacts'?
- How to determine the weights or importance measures of the different elements before they can be aggregated? Exogenously or endogenously? If the latter is accepted, what data do we make use?
- How do we make the different elements summable by appropriate standardization?
- Do we take the absolute values or relative changes over the corresponding values in the previous stage in the chain of innovation?
- Should we calculate ideality in the same fashion for all innovative products or use different procedures for different product categories?
- How do handle the case of innovative services?

Further, to examine the law of increasing ideality in the context of TRIZ, we may plot ideality versus number of innovation stages and possibly find that a logistic curve gives a reasonably good fit. It becomes pretty difficult to identify and, subsequently, to quantify the various costs and benefits of the innovative product or service. This is particularly so when we try to delve into the future or the life cycle of the product or service. And we may not have any scope to bring in the costs and benefits of any similar or analogous product or service, at least in all cases.

While quantification becomes necessary for comparison and, hence, for an assessment of progress towards the ideal, we should not be bogged down by problems of measuring desirable and undesirable effects of any attempt to improve upon the existing entity (technical system). The focus could first be on the elimination of or, at least, reduction in the costs or other undesired effects. At the same time, we can think of function augmentation and/ or use of ideal substance, ideal form and ideal process. As illustrated by Petrov and Seredinski, hulls of modern submarines are produced from titanium, because of its corrosion resistance property in sea water, absence of magnetic properties and conducive mechanical properties. Many parts of automobiles are now made from magnesium and the body from fibre glass. Altshuller mentions that 'metal parts work only on stretching, wooden parts only on compression.' For underwater applications, the ideal form is a sphere because it has high stability and density and has minimal relations between surface area and volume.

All this said, we should keep in mind the problem of 'measuring' ideality with the aim of coming up with a more convenient quantification of techno-economic progress, which can be used to examine the effort-worthiness of an 'innovation'.

9.4.2 Fundamentals of TRIZ

The eight trends of classical TRIZ are as follows:
* Increasing ideality (sum of benefits divided by costs and harmful effects)

- Stages of evolution (sigmoid curves)
- Decreased human involvement
- Non-uniform development of parts
- Increased complexity followed by simplicity
- Transition to micro-levels and increased use of fields, for example, optical/magnetic fields
- Increased dynamism and controllability
- Matching/mismatching of parts

TRIZ is not a black box but does transform problem analysis into problem-solving.

In TRIZ, we first recognize some 'technical' contradictions—calling for 'trade-off' or compromise between two features where improvement in one is associated with degradation in the other—and some 'physical' or 'inherent' contradictions where some feature or even component is 'required' as well as 'not required', at least at the same time.

Altshuller started with a 39 × 39 matrix of contradictions relating to engineering features of a 'technical' system. He then proceeds with 40 inventive principles and indicates which principle(s) can be used to remove a particular contradiction. Of course, really imaginative or out-of-box thinking may be required to eliminate an inherent contradiction.

It must be remembered that a new or an innovative product or service developed now—with or without involving TRIZ explicitly—may appear to be just simplistic and nothing novel, once it enters the market and taken up in the society.

Innovation is an ongoing exercise towards achieving the 'ideal final result' and at each stage, the ideality increases. Altshuller is credited with the development and implementation of a theory that really deals with innovation in terms of resolving a few contradictions—physical as well as technical—present in quite a few existing products or 'technical systems', in the words of Altshuller. Let us consider some of the 40 inventive principles and indicate their application in certain social systems to increase their 'progress towards ideality', or broadly speaking to enhance organizational performance. Some of the principles are described below:

*Principle 1: Segmentation—Divide a subject to be taught/discussed into naturally/logically distinguishable units.
*Principle 2: Taking out—Using the barking sound of a dog for a burglar alarm.
*Principle 4: Asymmetry—making a shaft non-cylindrical to avoid slippage.
*Principle 6: Universality—Child's stroller also a shopping cart. A tooth-brush also a tongue-cleaner or a hammer also a nail-puller.
*Principle 13: Doing the other way round—meeting a worker at his workplace for a discussion, instead of calling him in. A more appealing example is making the static part of a system dynamic and vice versa, for example, an elevator where the passenger remains idle while the staircase moves
*Principle 16: Partial or excessive action—Give a student only a partial solution to encourage him/her to do the rest on his/her own.
*Principle 19: Periodic action—Send frequent reminders to get work done.
*Principle 23: Feedback—Automated statistical process control.
*Principle 24: Intermediary—Join glass to metal by a series of materials to achieve change in thermal expansion in steps.
*Principle 25: Self-service—Use furnace flue gas to pre-heat incoming fuel gas.
*Principle 26: Copying—Use aerial photographs to count sheep, measure land areas, identify crops grown, etc.
*Principle 33: Homogeneity—Cut diamonds using diamonds.
*Principle 39: Inert atmosphere—preserve food in packets emptied of air and/or filled with nitrogen.

Having examined a huge number of patents, Altshuller came out with 76 standard solutions, which can be used as a database for possible use by an innovator or someone engaged in designing and developing some new product or service, to meet some existing but unmet needs or new and additional needs or even imagined or inarticulate needs.

9.5 ARIZ

Apart from formulating the inventive principles to remove the inherent contradictions that hinder the achievement of the ideal final results, Altshuller also proposed an algorithm for problem-solving known by the acronym ARIZ, which has undergone many revisions starting from the initial version ARIZ 56.

At the beginning, the intention for developing ARIZ was inventing a set of rules for solving a problem, where solving a problem means finding and resolving a technical contradiction or, for any given solution, ensuring less material, energy, spare and time used. The method was intended to include typical innovation principles, such as segmentation, integration, inversion, changing the aggregate state, replacing a mechanical system with a chemical system.

- ARIZ 56 is a set of steps based on the practices of the best inventors. It also incorporated the idea of thinking beyond the boundaries of the immediate subject.
- ARIZ 61 had three stages.
- I. Analytical stage
- II. Operative stage
- III. Synthetic stage

Analytical stage

- Step one
  State the problem. This is the most important step. Problem has to be stated without jargons.
- Step two
  Imagine the ideal final result (IFR). At this stage, it is not important whether the problem can be solved as per the ideal final result. An insight of IFR will give us the ultimate goal to be achieved.
- Step three
  Determine what interferes with achieving this result, that is, determine the contradiction that exists.
- Step four
  Determine why it interferes, that is, the reason for the contradiction.
- Step five
  Determine under what conditions it will not interfere, that is, by doing what the contradictions can be removed.

Operation Stage

This stage is the main problem-solving stage. The following steps elaborate the guidelines of this stage:

Step I: What are the possibilities of making changes in the object or system or process technology?

a. Can the size be changed?
b. Can the shape be changed?
c. Can the material be changed?
d. Can the speed be changed?

e. Can the colour be changed?
f. Can the frequency be changed?
g. Can the pressure be changed?
h. Can the temperature be changed?
i. Can the working conditions be changed?
j. Can the time of operation be changed?

These are general questions. They are indicative and not exhaustive. The central idea in this step is to think of what will happen, if this is changed?

Step 2: Explore the possibility of dividing an object into parts.
a. Identify and isolate a weak part.
b. Isolate the crucial parts which are necessary.
c. Identify identical parts and separate them.

Step 3: Is there any possibility of altering the outside environment?
a. Can any parameter of the environment be altered?
b. Can an environmental parameter be replaced?
c. Can environmental factors be separated into several media?
d. Can any gainful function be achieved by environmental parameters?
By this step, we are trying to exploit any environmental resources or energy that can be put to productive use.

Step 4: Is there any possibility of making changes in the neighbouring object?
a. Examine whether there is any relation between independent objects participating in the same function.
b. Examine whether we can eliminate one operation and transfer its function to another object.
c. Examine if number of objects can be operated in parallel on a defined area.
By this step we try to assess the interaction between independent objects and try to eliminate or transfer this or to do both.

Step 5: Examine whether a similar problem has been solved in another industry, by any another technology, by anybody else, anywhere.
Step 6: Come back to the original problem. If any of the above steps could result in a solution well and good. Otherwise, redefine the problem in a more general way and re-examine.

The synthetic stage

- This stage is the implementation stage.
- Step 1: Try and change the shape of the object. It is normal for a new machine with a new function to have a new shape.
- Step 2: If some corresponding or interacting object or a system has to be changed or modified this has to be done.
- Step 3: Examine whether such a solution is applicable to any other object or function of the system.

ARIZ 85 – B and C

- Significant structural changes were made.
- The link between the algorithm, the system of standard solutions and patterns of technological evolution became stronger.
- Up to this time, Altshuller was concentrating on the efforts of ARIZ improvement. But he started ignoring suggestions saying ARIZ 85 C is adequate and started concentrating on the theory of development of a strong creative personality (TRTL) ignoring ARIZ.
- As can be seen, ARIZ has been evolving and getting modified by the inputs from recent practitioners in the Western and European world.

9.6 INSTITUTIONALIZATION OF INNOVATION

The role of creativity and even of fantasy was recognized by Altshuller. Of course, Altshuller did not focus on talent–creativity–innovation inter-relationship explicitly. It has been claimed by some authors that creativity results from expertise, motivation and imaginative thinking. It is true that creativity is an inherent ability or trait of an individual, while innovation is a creation usually by a team. It is felt that talented students can easily pick up TRIZ principles and can come with novel solutions to problems awaiting solutions or, in some cases, solutions better that the currently available solutions. TRIZ duly recognizes the role of out-of-box thinking or lateral thinking.

Attempts should be made to take TRIZ principles as well as other principles for thinking out-of-the-box to students, especially the talented ones, to nurture inherent capabilities for innovation. This is in the context of students who are talented, but do not excel in innovation necessarily.

Creativity could be of different kinds. At the highest level, it is the birth of a new idea or a theory which is qualitatively different from and not conforming to or deducible from any existing paradigm, and which explains a wider set of natural phenomena than any existing theory. There is creativity of another kind at a different level, of a discovery made within the framework of an existing paradigm but of immense significance in a particular discipline. Both are indeed sources of new knowledge. However, there is a subtle distinction; in the first case, it is created a priori though confirmed later by observed facts, and in the second case it is a logical extension of current knowledge.

No discovery of great importance has ever been made by logical deduction or by strengthening the observational basis. It is then clear that a necessary condition for creativity is to let the mind wander unfettered by the rigidities of accepted knowledge or conventional rules. Perhaps, the thinking that precedes a discovery is of a fuzzy type, a successful interplay of random search for new frameworks to fit past experience and subconscious reasoning to narrow down the range of possibilities. Describing the act of creation, Arthur Koestler (*The Act of Creation*, 1990) says,

> At the decisive stage of discovery the codes of disciplined reasoning are suspended as they are in the dream, the reverie, the manic flights of thought, when the stream of ideation is free to drift, by its own emotional gravity, as it were, in an apparently lawless fashion.

Random thinking is an important ingredient of creativity. Other elements are required, such as the preparedness of the mind, ability to identify important and significant problems, quick perception of what ideas can lead to fruitful results and, above all, certain amount of confidence to pursue difficult problems. Hofstadter says random thinking is an intrinsic feature of human thought, not something which has to be artificially infused.

In the context of business organizations including academic institutions and R&D establishments run by business houses, corporate leaders and leaders in the human resource management function have to play significant roles in (1) mapping creativity and free-thinking capabilities of employees, irrespective of their qualifications and experiences, job placements and performance levels currently to identify individuals who can be involved in the organization-wide innovation exercise or in new product development team, maybe with some exposure to enhance their current capabilities to accept challenging jobs and thinking out-of-the -box; (2) recruiting a few talented persons with creative capabilities—proven or even potential—who are willing to take up a new product development effort by himself (herself) alone or to join or even lead a team for innovation; (3) retaining talent and creative capabilities through creation and maintenance of a supporting environment that allows flexibility and freedom (within some well-thought out boundaries), provides resources necessary to carry out trials and experiments besides procuring relevant data and documents and appreciating and rewarding—individually or collectively, depending on the situation—any new, concrete and actionable idea or design or model and innovative academic work reported in some prestigious journal or by a reputed academy or society in the concerned subject area.

10 Technological Innovation

10.1 ASPECTS OF TECHNOLOGICAL INNOVATION

In the previous chapter, how innovation can be fostered using the TRIZ concept and new product development were discussed. We should note that not all innovations can be justifiably called technological innovation even if it involves a recent technology. New and novel products and services can be designed, manufactured, tested and marketed employing the efforts illustrated in the previous chapter, and such efforts need not be linked to a new technology. In fact, sustaining or incremental innovations that helps in adding new or novel features or functionalities, or both, to existing products and services meant for existing or new markets do not involve any new technology in many cases. Even process innovation involved in designing and/or generating and/or delivering some existing product or service—expected to bring about substantial improvement in business performance—may not necessarily involve a new technology. Innovation, which leads to a structural change or a shift to a different business model, may yield benefits to an organization which still employs its existing resources, including processes and staff. A step ahead of continuing with existing technology and still working out innovation would be to use a recent technology, with suitable adaptation, if required. Without technological innovation as we consider in the following, simply innovation without the development or deployment of a new technology will not help an organization to create a better value proposition and to sustain a superior level of performance, compared to others in the fray.

Technological innovation can be defined as a process of creating or using a new technology to come up with new or improved products and services that will be market-ready, to significantly change the way people meet their needs or carry out their activities. The second part of the deployment of technological innovation points to its benefits to people in both production and consumption of goods and services. Beyond the use or deployment of new technologies, technological innovation does cover the more engaging and rewarding activity of technology development or creation, validation and deployment (by self or others) of new technologies. It is generally argued that the development of new technologies is meant to be primary activity of research and development establishments and even of specialized institutions of higher education in engineering, pharmaceuticals or management. Instances do exist where a manufacturing organization engaged in mineral extraction,

DOI: 10.1201/9781003591733-10

drug manufacture or production of some building material has come up with a new technology. Incidentally, technological innovation subsumes digital innovation in all its ramifications.

Two different aspects of technological innovation emerge: the first method is developing new technologies and carrying out necessary tasks to render any new technology deployable and the method is concerned with a meaningful absorption of a newly developed technology after modifications, if needed, to come up with value-creating or value-adding products and services. Further, innovation can be information-based (called 'soft' innovation) or material-based (called 'hard' innovation). Soft innovation primarily refers to digital innovation. We have to keep in mind that digital innovation has a wide ambit to include all types of organizations which are engaged in either development or absorption or both of new technologies. It will not be unfair to claim a much bigger role for digital innovation (that embraces digital transformation of the organization) compared to those of technology development or technology absorption, referring basically to technologies for design, production, control and similar activities. Digital innovation may change the very business model currently in use and result in a radical improvement in organizational performance. Digital innovation may result in a change in organizational structure and management practices—job placements, communication and control systems, interaction mechanisms relevant to external agencies like regulatory bodies as well as customers and investors. In terms of scope of application, digital innovation has an edge over technological innovation (involving hard technologies), which is confined to specific products or processes. In sum, technological innovation generates new technical knowledge or/and improved use of available knowledge.

Pointing to the impact of technological innovation, we may note that at the meta-level, such innovation (1) modifies the structure of industries in terms of product and service profiles, (2) changes the composition of demand in the labour market in terms of skill requirements, (3) alters the competitive positions of nations, (4) stimulates economic growth and (5) increases the well-being of people in the society. At the micro- or the organization level, technological innovation (1) affects the competitiveness of business organizations and (2) gives an orientation to organizations in designing their strategies. To retain its position of performance, technological innovation becomes an absolute necessity for any organization. Since marketing practices may undergo changes (taking advantage of new technology) and even sourcing of finance may not remain the same, it will be wise for any organization to work out a portfolio of innovation strategies from which the innovation that fits into the changed scenario should be chosen for adoption or development.

We will take up the case of technology development first, referring to this as one of the main activities of an R&D organization (usually taken up on a project mode) and as a desirable activity of an academic institution imparting higher education in a specialized field. This will be followed up by a discussion on technology adoption by any organization. Steps involved in each of these cases will be briefly outlined. The aspect of digital transformation will be touched upon and the role of artificial intelligence will also be covered.

10.2 CANVAS OF TECHNOLOGICAL INNOVATION

To take a somewhat realistic and maybe simplistic view that innovation is all about problem-solving and there are many ways to solve a problem that any organization faces in its journey. Some problems are well-defined but may have a very difficult solution, not so easy to find out, while there could be some other problems which are ill-defined or defined in fuzzy or equivocal terms, which can be possibly solved easier by even a known method or technique. All this has resulted in a two-way classification of innovations based on how it can be defined and the domain of its potential applications (Kelley and Kelley, 2014). The four classes or types that emerge are usually known as (1) sustaining, (2) breakthrough, (3) basic (research) and (4) disruptive. It should be added here that sustaining innovation may be differentiated from incremental innovation, which is actually the outcome of continuous improvement usually at the shop-floor level by small groups of people working on the problem. Incremental innovation does not generally involve any change in hardware, while this change is usually called for in sustaining innovation. In the communication industry, advent of the smartphone was a breakthrough innovation bringing in a new technology. The subsequent updated versions with added functionalities through updated hardware illustrate sustaining innovation, while regular updates involving improved software to benefit the existing users can be called incremental innovations.

Innovation can also be classified as radical, incremental and architectural. In architectural innovation, knowledge components and core concepts underlying them remain the same, but knowledge about linkages among the components has changed. Somewhat similar is the approach to recognize product innovation, process innovation and service innovation as technical innovation forms. Administrative innovation and some forms of service innovation are not to be regarded as 'technical' and may or may not be required for technical innovation.

Another interesting dichotomy arises between competence-enhancing and competence-destroying innovations. In the first one, innovation builds upon existing knowledge and skills within the organization. It is incremental in nature and helps the organization to leverage their existing competencies to improve their products, processes or services. A good example is provided by the automobile industry in which over the years incremental innovations have resulted in better fuel efficiency, greater comfort and higher security and safety. On the other hand, competence-destroying innovations involve a radical departure from the existing processes and competencies to the extent of even rendering them as obsolete. A typical example is the emergence of digital photography that has rendered the obsolete the set and equipment used for traditional film-based photography.

Kelley (2014) speaks about eight core design abilities that the innovation team charged with developing a new technology in a corporate environment should possess and deploy in order to design the course of innovation. These are (1) navigate ambiguity, (2) learn from others (people and contexts), (3) build and craft things intentionally, (4) communicate deliberately, (5) design or plan your 'design work', (6) synthesize information, (7) experiment rapidly and (8) move between concrete and abstract. This list is not to be taken as a general prescription and has to be suitably modified to fit into the specifics of the problem and to the organization facing it.

10.3 TECHNOLOGY DEVELOPMENT

The development of a new technology owes its primary motivation to advances in basic scientific knowledge that are taking place in academic institutions and research organizations. However, new technologies that are developed in this mode may not necessarily address emerging problems in business organizations producing and providing hard and soft goods. This explains why some of the business organizations sponsor research projects for organizations specializing in R&D work to develop technologies to suit their needs. Of course, technologies that emerge from projects based on recent developments in science may also open up avenues for the creation of new designs, processes, equipment, products and services. Thus, the two outcomes of technology development are (1) technological push coming from scientific research and development sector, with no necessary commercial purpose at least in the short run, and (2) market pull wherein market needs instigate organizations to develop new technology that can satisfy the demands of consumers and businesses. In the second case, organizations can absorb and deploy technological innovation like digital transformation making use of their competencies and putting in efforts towards the development of new technologies.

The steps involved in technology development involve, by and large, basic research and more importantly applied research that can lead to the development of new technologies or even some output that can be followed up with the help of some experts to result in a full-fledged technology, which will be application-ready. Development work to create prototypes of a product or a service, followed by due testing of the outputs for their usability, scalability from laboratory scale or pilot-plant scale to commercial production scale and then marketability of the resultant product or service. A demonstration to establish novelty or superiority of the technology and the product or service can be created through experiments and examples .The newly developed technology can be made available to organizations intending to absorb it, by identifying the requisite equipment and manpower and providing support by way of guidance and troubleshooting during absorption of the technology with or without necessary modifications by any client organization. New technologies may also be developed to solve certain existing problems in designing or producing some product or service already identified as potentially beneficial to customers and users or to develop concrete ideas about the felt needs of people in general—in the absence of some fancied or imagined products or the absence of some desired features and functionalities in some existing products—which have remained unmet and to come up with some new or improved products or services that will be immediately taken up in the marketplace. This is a stronger and more common motivation for the development of new technologies.

Even in this context, a planned sequence of activities is involved rather than some hit-and-miss or trial-and-error activities. The following sequence has been recommended by several experts in technology development by R&D organizations as well as by organizations normally engaged in production processes.

- Find an unmet need of users as well as developers, people engaged in manufacture or in maintenance of existing products, which needs be improved upon, or when the needs can be met by designing, producing and marketing of some new product or service.

- Collect information about requirements from stakeholders relating to availability of resources to work out tentative solutions in terms of broad outlines of design, manufacturing or production process, testing and validation of the tentative solution.
- Work out some possible solutions and find out to what extent the unmet need can be potentially met through any of the proposed solutions.
- Put the most plausible and workable solution to design the new features and functions or to develop a new product or service altogether and build up a plan for validating and verifying the design or for manufacturing the new product or creating the new service, in consultation with concerned stakeholders.
- Build a few prototypes for the purpose of testing and subsequently, if found in order, release it for test-marketing in order to get feedback about the realized performance of the product or service under consideration against expectations and requirements. This is likely to help in identifying the needs in clear and concrete terms.
- Release the first batch on a commercial scale after incorporating changes suggested in the previous step and set a mechanism for effectively 'listening to the voice of the customer' and start with quality function deployment exercise based on a proper analysis of customer requirements, including legal, environmental or social requirements imposed on the organization or accepted by the organization in its policy.
- Communicate with existing as well as potential customers in clear and convincing terms and with proper content to tell them about your value proposition and why should people like the improved or new products and services, avoiding any dishonest claims.
- Convince stakeholders about resulting benefits to the organizations and the need to continue with such activities through adequate training for the concerned workforce, motivating workers to think creatively, acknowledging and even rewarding ideas about technological innovation, including absorption and use of a recent technology that can help works in a certain area to facilitate and improve their performance.

10.4 ABSORPTION OF NEW TECHNOLOGY

Organizations learn, accept, adapt and use innovative technologies to pursue the course of consistently improving their performance, even if they cannot by themselves develop new technologies. Technology adoption is affected by performance expectancy, effort expectancy, facilitating conditions and social influence. A for-profit organization has to necessarily weigh the anticipated effort and the time and cost consequences thereof in adopting a new technology against the expected net gain by absorbing and making good use of the new technology. However, the first part is more related to the present, while the gain may be distributed over a long time horizon. Some organizations may find its leaders and its workforce as welcoming new ideas, technologies and methods of work and ready to undergo any training to embrace a new technology and such a situation should encourage such organizations to go in for new technologies and to make the best possible use of them. Situations may also exist where social considerations, including environmental issues, may push local organizations to adopt currently available environment-friendly or otherwise beneficial innovative technology. There could exist other factors to encourage forward-looking organizations to seek, learn, plan to accept, adapt to own situations and contexts, motivate leaders and secure their commitment, ensure willing involvement of the workforce and mobilize resources needed to successfully implement new technologies in their business processes.

Technology adoption is not just buying a new technology and its ready use. A new technology developed in a different context by a different (maybe R & D) organization cannot be grafted to the processes in another organization. Within the organization opting for technology adoption, there will some segments who readily learn and even modify the new technology to suit the processes and the equipment they handle, there will be others who will absorb the new stuff without any hurdle and still others called 'laggards' who will trail behind others in accepting the new technology without any reservation and agreeing to put in efforts to implement new processes and work practices. For the technological innovation to provide real gains in performance and results, it has to be effectively and efficiently put into practice in the entire organization, which may have varying impacts on the current practices in different segments or functions.

Technology adoption to enhance organizational performance has to be necessarily a corporate function, spanning several steps in a sequence and taking adequate care to ensure that the projected adoption can be completed without any negative or undesired impact on the existing basics of organizational activities. Some of the steps have been cited in literature and are as follows:

- *Technology selection* based on a careful analysis of business needs, including growth and improved performance, on the one hand, and resolution of current business problems, on the other. Usually, several new technologies may be available, though knowing the choices may imply interactions with external experts or internal exercises by motivated and innovation-oriented staff members. Once choices are known, an important corporate activity would be to evaluate the feasible choices and to select one as the most promising based on a resource-view approach. As indicated earlier, some competence-enhancing innovation with reasonable expected benefits and low risk may be preferred.
- *Planning*: Management has to review its infrastructure, staff, strength and weakness, and existing job allocation and work practices to identify the nature and extent of changes required to absorb the new technology. Integrating a new product technology involving a new process calls for a greater resource mobilization than a new product technology based primarily on existing processes. Adoption and integration of a new process technology may be a challenging task and may involve some sort a business process re-engineering. Resources do not just mean new equipment, new shopfloor layout and others. The more important resource to be procured and deployed could be in terms of competent and motivated workforce, which is ready to pick up the new system and signalling a departure from previous marketing approaches and highlighting the benefits to be derived by customers.
- *Communication*: A key component of the plan for adoption is ongoing communication encompassing the entire workforce and some business partners like those in the supply chain to convey in clear and easy-to-understand language what all is going to happen, why and how the new technology will be a win–win situation

for the organization as well as the people concerned. This must start right from the very beginning so that people are not taken by surprise once the decision to implement a new technology is made without keeping them in the loop. Surprise may give rise to suspicion, withdrawal, denial and even resistance. Leadership has to upscale this activity to convince middle management about the need for adopting an innovation and the benefits likely to accrue to both the organization and the entire workforce, involving them to spread the message correctly to people working under their guidance. In the process, special attention may be given to those who champion the cause of innovation adoption.

• *Training and motivation*: Another key element in planning is to map the needs of training for people at different levels and tasked with different activities to familiarize them with the novel features they might not already be knowledgeable about. People also have to be motivated by a due explanation of the benefits in terms of conveniences in working and improving individual performance that may lead to some awards or monetary gains. In fact, early on any gainful result accruing from adoption of an innovation should be celebrated to felicitate all in the organization. It has been sometimes noticed that inherently competent middle-level executives prefer status quo with a little tinkering, which they can inject in the functions under their control. This category of people should be taken into confidence first and be convinced about the need for embracing the selected innovation, so that they themselves can become champions. In fact, they could better be involved right at the time of deciding to for some innovation and in selecting some type of innovation.

• *Testing and development*: This is done to check if the new technology can work without any hassle in a few selected areas and if some minor adjustments or modifications need to be incorporated before full-scale absorption of the selected technology. Quite likely is the need to make small adjustments in some procedures or to fine-tune some others. This may require an internal expert or a consultant hired to fix the nitty-gritty before the technology adoption is integrated throughout the organization.

• *Resource mobilization*: Except for a publicly funded organization meant to provide services to identified beneficiaries as a matter of public policy with a commitment irrespective of the financial implications, procurement of resources needed for the prospective absorption of an innovation may pose serious problems to the corporates. A general advice may be to avoid procuring the entire resource requirements at one go and to phase out the mobilization process carefully by a scrupulous scrutiny of the timelines for deploying the resources. This problem may be compounded by even a remote possibility of the new technology failing to yield the expected benefits, especially in terms of financial gains.

• *Monitoring and evaluation*: As the new technology starts getting integrated in the business processes of the organization, top leadership has to monitor the progress of integration and the benefits in various forms—both financial and non-financial—that are being derived from it. Several metrics for technology adoption have been suggested for monitoring and evaluating this progress and to initiate appropriate steps for setting the project of technology adoption on the right track.

10.5 TECHNOLOGY ADOPTION LEADING TO INNOVATION

Tracer studies have revealed that the successful integration of a new process or product technology in a few organizations have resulted in the development of a new technology to create a novel product or initiatives to develop a new technology for an existing or a new product incorporating better features and functionalities. In fact, this is often regarded as a downstream effect of a successful adoption of innovation. This explains how several useful innovations have emerged from organizations not primarily or exclusively engaged in R&D activities. Of course, academic institutions which embrace new technologies for improving the teaching–learning process also come up with many innovations.

One explanation for this phenomenon lies in the fact that ‘nothing succeeds like success’ and people involved in accepting, absorbing and taking due advantage of some innovative technologies with a zeal and deriving expected benefits feel enthusiastic about the constituents of the new technology. The innovation champion and even some members of the innovation team take up exercises like reverse engineering to understand how to incorporate elements in a design or an architecture that can bring about new and non-trivial features in a process or a product. They then try out their hand in experimenting with some innovative ideas that can result in either a new design or a new process. And, in some cases at least, such enterprising members of the innovation team do succeed in developing an innovative product or process that is likely to be highly accepted in the market. The team should also examine the feasibility of building prototypes and establishing market-readiness of any innovation under their consideration. Top management may allow some test marketing with their in-house innovation and provide the internal team with an effective feedback from potential customers. The team may take due account of the feedback and modify the innovative process or product to remove any deficiencies.

From a technology adopter to a technology developer is indeed a big leap forward and a welcome step in the journey towards excellence. It is not completely wrong to assert that an organization that already deploys appropriate technology and also has the capacity to absorb new technologies effectively and efficiently will find it difficult some day to catch up with some and will be left behind them in the desired pursuit of excellence.

10.6 ROLE OF ARTIFICIAL INTELLIGENCE

One of the most important roles that AI can play—if properly deployed—is in decision-making. AI can be a very effective instrument to assist a decision-maker—especially in a complex situation—and can even take up the role of a decision-maker in situations in which it can surpass available human intelligence. In fact, through the development of a large data platform, machine learning to process large volumes of data quickly and to provide insights into the future and using large language models to comprehend human interfaces, AI can contribute to decision-making at three levels—decision support, decision augmentation and decision automation—depending on the existing human resource capability to work in tandem with AI.

Decision support systems were discussed much earlier when organizations would lay stress on data-based decisions, develop and maintain a database for the purpose, drawing upon internal data from various areas and procuring data from outside about other organizations, markets and regulatory mechanisms in place. Relevant data would be processed using appropriate software and descriptive as well as predictive analytics applied to the data would provide important inputs for decision-making by concerned executives. There were then limitations on data and the extent of data analysis. More efficient software was gradually developed and decision support systems are becoming more useful.

Beyond supporting decision-making by humans, AI took up the role of expanding the database to work upon by accessing data—even unstructured ones—from different platforms and building up a big data architecture to facilitate application of many efficient techniques for prediction and for optimization. This way, AI can offer many more alternative solutions to the decision problem than a human can identify and offer. In this case, prescriptive analytics will be taken advantage of. Further, in the context of decision-making under uncertainty and 'deep uncertainty', AI can help in generating many possible scenarios for the future, which would not have otherwise occurred to the decision-makers nor would have been suggested from a hard look at the past. Decision automation may not be accepted, except in situations where consequences of a wrong or inefficient decision can be controlled suitably at an appropriate later stage. Steyvers and Kumar (2023) talk about challenges to AI-assisted decision-making, in terms of preparing the organization to make full use of AI in decision-making.

Effective communication in terms of the right content at the right time and in the right manner is, indeed, a sine qua non for any well-performing organization. In certain operational areas like marketing, the enormity of the frequency of communications and the number of recipients involved makes it impossible for the people in charge of such communications to do justice to the task of fully utilizing the implications of the communications. (Tele-callers engaged by a few organizations for sales promotion purposes are usually poor communicators and are not being considered here.) Marketing is a broad area of great interest and to proceed with some concrete and useful deployment of artificial intelligence, let us confine ourselves for the time being to customer feedback only (including communications to and from prospective customers). This feedback does convey complaints (ranging from delayed response to requests regarding time and mode of delivery to receipt of products and services of poor quality), suggestions (mostly unwarranted) for possible improvement in goods and services and requests for repair and replacement. Even simple queries about product/service features demand timely and proper response. It is quite expected that many of these requests or queries, suggestions or complaints crop up repeatedly. These can well be treated as important sources of data for analysis of quality costs.

AI can be very effectively used in the above context firstly to acknowledge receipt of any such communication with a response in a standardized content almost immediately on receipt. AI can then segregate the incoming communication into complaints, suggestions, queries and requests. Classification may be decided by the organization. AI—in its simplest use—can be used to forward the communication content to the concerned executives. In fact, AI can even carry out a detailed clustering of the complaints (or the suggestions or requests) into meaningful clusters based on technical as well as financial implications. AI can also convey to the concerned person(s) the likely time to be taken for a satisfactory resolution of the problem pointed out. For the organization to prioritize its remedial responses and actions, AI can be used to classify the concerned customers based on their volumes and frequencies of transactions with the organization.

By making effective use of AI, employee time for routine, repetitive tasks can be totally avoided and freed for use in more important areas. Beyond releasing employee time for more effective purposes, the routine and repetitive tasks left to the care of AI will be performed without delay, avoiding silly mistakes and with conviction to the customer or any other interested person. As mentioned above, AI can track time and cost involved in resolving any dispute or in executing a repair or replacement action within the declared warranty period and eventually prepare a report on the analysis of cost of poor quality. AI can extract any useful actionable suggestion received from a customer, a regulatory agency or a custodian of public interest. ChatGPT has been and can be gainfully used to come up with promotional material and patterns of display or dissemination.

Secondly, AI can bring about improvements in communication and consequent action affecting people within an organization, such as convening a meeting by an executive. This may be an apparently innocent and trivial task, but trivialities have considerable affective potential and the latter can be linked to its impact on productivity. One simple case is where an employee applies online for leave, help or support from his superior and the latter delays action on the same willy-nilly, possibly in the face of some more pressing work on the table. All that is needed is a simple nudge or a reminder to look up the application and take the required action. It is just possible to extend the situation to one where a decision or an action has to be taken in a meeting and somehow convening the meeting gets delayed. AI can easily act to convene the meeting, ascertaining the availability of the members to be involved.

Let us consider the critical area of quality assurance, combining quality control during production and inspection, testing and remedial action during and immediately after production. Deviations detected during production can be instantly notified to the operator and even acted upon autonomously if AI has access to the whole array of corrective actions corresponding to the nature and extent of deviations detected. Taking advantage of AI for efficient execution of training and induction programmes has been widely acknowledged.

11 Managing Change

11.1 IT'S ALL ABOUT CHANGE

Unless we change the way we look at others, or even at ourselves, in respect of our needs and the manner in which we like those to be met, any discussion on organizational excellence may be just sound and fury. We all want improvements in our activities and expect improvement in outcomes from those activities. Improvement necessarily involves change. In fact, improvement is, in a simplistic sense, change for the better. Of course, changes may lead to a worsening of the present situation as well. This fact points to the need for designing changes appropriately so as to remove current bottlenecks in the way to achieving the desired levels of performance and to pave the way for superior performance. It will be quite wrong to equate unsatisfactory organizational performance (sometimes threatening the very survival of the organization) to a stagnation in its architecture, system or processes and activities. Rather, we all find changes taking place in all these characteristics of different organizations—some struggling to survive, some thriving and growing and a few doing exceptionally well. Going a little beyond, we also notice changes—quite a few being significant with a potential to destabilize extant systems—in the environment within which organizations operate. These external changes are concomitant with advances in human knowledge impacting on human behaviour and the changing profile of offerings from organizations in the manufacturing and service sectors. Some organizations are simply swept away by these changes without being able to turn the same to their advantage, while others foresee the need for change and come up with a proper response to changes in the external environment that are likely to impact organizational performance.

Change is known to be the only 'constant' in life, taking place in every aspect of any organic or inorganic system, silently or noisily, visibly or in a covert manner, creating a ripple or a wild turbulence, on a continuing basis (allowing breaks). 'The old order changes, yielding place to the new' is not just a proverb, but a hard reality. Individuals change their competencies and behaviour as well in their needs and aspirations, markets change to contactless e-markets, products and services change in their features and functionalities as well as in their price and quality, communication practices change to involve faceless responder, and so on. Organizations change inevitably.

 DOI: 10.1201/9781003591733-11

Climate and culture are two dominant determinants of organizational performance. Both the constructs are conceptually complex and have diverse definitions.

In this context of change affecting outcomes and impacting whatever individuals and organizations do, we also come across a whole lot of differences in the end results, ranging from 'disastrous' through 'amenable to recovery' to 'boosting operational performance'. Such differences can be attributed to the extent changes are 'managed' by various organizations. In fact, the history of managing change initiatives has not been a story of only successes. While the situation has improved over the years with better planning to implement change initiatives, the findings of a Harvard Business School from a study during 1980 and 1995 showed that only 30 per cent of change efforts produced an improvement in bottom line results. Even in some Fortune 500 companies, results were no better. However, the uptake of change initiatives has grown considerably in both profit-oriented and in not-for -profit organizations, including public service organizations. We have instances of organizations in various sectors where visible and impressive improvements achieved have been quite inspiring to others. Literature on organizational change is copious with several well-written books entirely devoted to the subject besides numerous articles published in journals and many interesting blogs. A periodical titled *Journal of Change Management* does contain highly academic articles including reviews, on the one hand, and very informative and interesting case study analyses, on the other. The book *Making Sense of Change Management* by Cameron and Green (2004) is a classic on the topic, not to be missed by a serious reader. Issues taken up by the authors on the subject relate to models and metaphors for organizations, types and motivations for change, leading or managing change, steps involved in dealing with change, impact on organizational performance, changes initiated in the service sector, besides a host of related aspects of change management. On some issues, we find a convergence of opinions while on a few others, we come across completely divergent views. In some cases, a closer look can possibly lead to a near-consensus of views expressed.

Two levels of change have been generally discussed: (1) fundamental or transformational, addressing the big picture issues like structure, culture/strategy, leadership and external environment, and (2) transitional or transactional, addressing issues like structure, management practices, systems, motivation and workplace environment. The conventional dichotomy of change as incremental (local) versus radical (breakthrough and organization-wide) is somewhat different, since even a transactional change may lead to a significant breakthrough improvement, and even a transformational change may unfortunately end up in only a local improvement. Leaving aside such dichotomies, an important differentiation should be made between organizational changes welcoming and embracing contemporary developments in science and technology to elevate operational performance, not under any pressure from externalities nor under any impending threat to its current level of performance, and changes forced under some compulsion or in the face of an impending trouble. The latter is an adaptation to a change in the external world. There exist cases where changes in regulatory frameworks mandate some changes in policies or strategies or in processes and people. This is more likely to happen in public service organizations. Organizations eyeing excellence generally adopt a strategy that can conveniently and effectively accommodate different types of change using a mechanism to sense the need for any such change at appropriate times. Intentional changes, duly planned and carefully implemented, constitute an important marker in the path towards organizational excellence.

In the pages that follow, we will discuss both adaptation to external changes affecting business of an organization and standing in the way of its desired performance trajectory as well as changes that are thought of and planned by leaders in an organization ahead of changes taking place in the external environment or not under any influence to change but with an urge to improve its performance significantly. This classification may apply along with the incremental versus radical improvement as the goal or objective behind a change. However, we will avoid discussing incremental changes on the shop-floor by people facing a problem, which they themselves can identify and resolve. In fact, change proposed for implementation in a department or a function involving people working in different operational areas will be covered here.

Conventional management and leadership typically issue instruction or call for action which directs or, at least, induces people at work to change in the way that the authority desires or requires. Conventional change management is a forceful intervention and is directed at the person or group from the change is sought. Conventional change management may accommodate big changes in one go. Managing such a change effectively to yield the desired outcome may be rendered more difficult in situations where people involved have fixed ideas and long-practiced ways of doing things and management—in a bid to quicken the introduction of change—tends to ignore or underplay the anticipated unwillingness or even resistance by the people concerned in accepting the change.

In this context, nudge theory proves to be useful as it typically an indirect approach which alters situations for people, so that choices are designed which produce options for helpful and voluntary change in people. Nudge theory, applied to change management, is designed to provide voluntary choices directed at the situation in which people currently exist and work. It usually involves an appropriate sequence of small changes being voluntarily accepted by the people concerned.

The role of change management may vary remarkably from one type of project to another. In the classical framework of projects being contracted out to some project-handling and managing organization working with people who can conveniently adapt themselves to the needs and demands of various projects, mostly in the infrastructure development and also in some plant or equipment installation and commissioning projects, the issue of change management does not pose as serious a problem as in the case of improvement projects taken up by a wide variety of organizations for implementation by themselves, may be with some support from some external agents.

In the case of improvement projects where an organization would like to adopt and customize a new process or a new system in the organization or some segment of it in order to improve its performance (measured in terms of both financial and non-financial results), making its own people willing and ready to accept changes associated with the project poses a different situation. Here again, if project selection is regarded as a component of project management (in a broad sense), then for the question of selecting among alternative processes or systems or even procedures for assessing improvement in performance, to be carried out by strategic management, change management plays a big role. In a sense, theory of constraints can be looked upon as a theory for change management in a wider context. The first two of the basic questions raised in ToC—What to change? and What to change to?—can be regarded as project selection while the third question—How to change?—corresponds to the project implementation component.

Several models for change management have been proposed over the years. Obviously, these have common elements, maybe in slightly different forms. At the same time, each model has some distinguishing features in terms of guiding principles or steps in implementation or expected outcomes. ADKAR is the acronym for a widely used model of change management with a focus on people management. The model begins with a meaning awareness—awareness about the present state of affairs (in respect of some indicators of performance and of results) and its deficiencies and hence the need for a change for the better and about the practice and prospects of a change in the base situation. D stands for desire to bring about the change, overcoming any inertia, any justification to maintain status quo or some anticipated problems. The third letter K is crucial for implementing change, which implies 'knowledge' about the nuts and bolts of the existing state, which are either impediments to improved performance or are creating setbacks in continued performance even at the base level. Knowledge is needed about the approach, the method and the actions to be taken at individual and collective levels to successfully implement the desired change. Is a priority area in change management and to equip the concerned workforce with requisite knowledge is an important responsibility of leadership. It must be remembered that unless change is desired by the people, it cannot be imposed on them. Once motivated to welcome the proposed change, people at all levels and in all tasks involved have to initiate appropriate actions (A). To ensure that such actions are proceeding in the right direction, reinforcements (R) are needed in the form of reviews, mid-term corrections whenever needed, appreciation and reward for success at tollgates, provisioning of adequate resources to be committed for implementing the desired change and sustaining the change once initiated.

11.2 DRIVERS OF CHANGE

While some experts distinguish between internal and external drivers of change, these two sets of factors are related one to the other and most of them can be traced to the world outside the organization. And as changes—especially those that create a lot of turbulence in the business environment—take place quite fast and in unpredictable ways, organizations are sometimes caught unawares and get into a muddle. This possibility calls for a continuous vigil on externalities and, based on their evaluation, organizations need to plan change initiatives on a regular basis, rather than being forced to change overnight, often unsuccessfully. Change initiatives are mostly meant to respond adequately and effectively to changes in the environment, though in most instances, change initiatives are taken up by organizations after taking into account their own desires and commitments to improve performance and to proceed faster in the journey towards excellence.

The drivers of change may include (1) change in customer behaviour—preferences, priorities and insistence—originating from their increased access to knowledge about new technologies and their adoption resulting in increasingly diverse product and service profiles; (2) change in market behaviour in terms of varied practices for pricing, offering discounts and freebies, stiff competition among distributors or agents of many organizations offering the same or similar products and services; (3) changes in regulations regarding occupational health and safety measures to be adopted, wastes discharged, environmental impact of different operations carried out by and for the organization; (4) changes in administrative controls on employment and wages etc. and, most importantly, (5) change in technology for production as well as delivery of goods and services. These changes have resulted in enhancing quality at the same cost, easier access and more convenience during usage, resulting in greater satisfaction to customers and other beneficiaries.

11.3 LEADING ORGANIZATIONAL CHANGE

John Kotter proposes eight steps for leading change in any organization:

- Create a sense of urgency, inspiring people to act with purpose and passion; build momentum
- Build a guiding coalition
- Formulate a strategic vision
- Enlist an array of volunteers throughout the organization
- Enable action by removing barriers
- Generate short-term wins or successes
- Sustain acceleration and
- Institute change, rather institutionalize change of behaviour and mindsets and ways of working.

The need for change in business processes and in organizational structure and behaviour should be felt internally at the appropriate time, based on a continuous evaluation of performance in the context of a changing environment, rather than imposed externally by a change in the regulatory framework or an economic slowdown or the emergence of new products and services threatening the acceptance of existing ones.

Three important tasks for change leadership include (1) communication of the need for change in terms of 'why?' and not 'what' or 'how' initially. Once the need for change is accepted or, better, welcomed by the people, then only the other two questions should be addressed. Now 'why' has to be communicated with substantive reasons based on facts and focusing on the consequences of not accepting the change on the performance of the organization and hence on the conditions of workers. Even the potential gain from the 'change' not only for the organization but also for the workforce should be duly emphasized. Skipping this step to exhort people in 'what' to change and 'how' to change as spelt out by the leadership will not bear fruit. The other two aspects related to 'what' and 'why' also deserve adequate explanation in clear and simple terms. (2) Collaboration of the workforce should be sought, since acceptance of change cannot be commanded. Unlike regular work wherein command and control may work in some cases, for change to be implemented successfully leaders have to seek collaboration and support from the workforce. (3) Commitment of leaders to bring about change has to be evident in all their activities, right from formulating a strategy for ushering in change in the processes to be identified, modalities of change and results to be expected by involving the workforce.

Change is easier said than accepted and implemented. Initially, people likely to be involved and influenced by the change, display some hesitation before accepting the change. In some cases, implementation of change gets thwarted by resistance from people even beyond non-co-operation. Willing and graceful acceptance of any change proposed is an absolute necessity for its successful implementation. All this makes change management a challenging task. Several approaches to change management have been proposed over the years.

11.4 ADAPTATION TO THE CHANGED ENVIRONMENT

For adaptation, Dervitsiotis (2007) identifies several phases:

- Sense the change creeping in the environment (consumer behaviour, production technologies, regulatory mechanisms, resource availability, investment climate, and the like) or the change that has already engulfed the externalities, through information flowing in or collected for the purpose (by way of business intelligence). An important task of top leadership is to plan and put in place an effective and efficient sensing mechanism, in which due advantage can be taken of Artificial Intelligence.
- Interpret the signals properly in terms of their likely impact on the organization's business. This requires a good amount of brainstorming with the concerned executives to get feedback about likely ramifications of the overall effect of the signals received on different functional areas.
- Decide on the nature, extent and areas of response to be made by way of changes to be effected internally so that any adverse impact on the internal processes can be warded off. Rather, the response can even be designed in a manner that beyond leaving out adverse effects, some positive impacts may be possible.
- Respond in the right manner, analysing the scenario revealed by the signals received. In case of the perceived change being rather small, an approach like adopting the Plan, Do, Check, Act cycle and/or fine-tuning the current organizational or process design, likely to result in incremental improvement may serve the purpose. However, if signals reveal a profound impact on the business landscape, the desired response may have to be system-oriented and pressed forward in several steps.
- Unfreeze the system by loosening bonds among different elements within an area and between areas and opening up these different elements to enable those to accommodate changes.
- Explore new possibilities in terms of new linkages connecting different elements of the system and even in terms of new elements created and/or existing elements dropped out, assigning responsibilities and equipping with authorities in new ways.

- Facilitate the emergence of a new organizational design, better aligned to the new business landscape and, finally,
- Guide change through adaptation principles.

All the above may imply some or all of the following:

- Setting up a team for scanning the emerging landscape and its dimensions as well as likely impacts
- Adding new nodes or taking out old ones like creating a new team for demand forecasting
- Adding new links or removing old ones to promote integration, consistent with independent functioning
- Adopting new rules for interaction for better co-ordination and fast response.

In this connection, let us recall the message of Lindblom and his 'squeaking wheel approach', sometimes referred to as the 'muddling through approach', to bring about some improvement in operational performance through the implementation of necessary change in the existing processes and procedures in a particular operational area. It may be argued that bringing about a radical change throughout the organization may lead to problems in implementation and the idea may have to be eventually dropped. It will be better to find out an area where some desired change can be conveniently worked out, because of people involved are already receptive to the idea of a change or even have suggested the introduction of some change by themselves. Lindblom suggests that organizational change should first be attempted and, hopefully, managed with success. A big story can then be made of the success achieved and the positive role played by the people involved along with the likely improvement in organizational performance. This will motivate other groups to come forward with suggested changes, which will generate worker co-operation required for smooth implementation. The choice of the place or group to start from may have to be determined by the top management by carefully listening to the 'noise' or disturbance being faced in the execution of assigned tasks, so that priority in attention can be paid to the most 'noisy' or problematic area. The rationale behind this approach is that a lock-stock-and-barrel change throughout the organization is quite likely to end in chaos, while small prioritized changes in chosen areas often result in genuine and perceivable improvements.

11.5 CHANGE MANAGEMENT MODELS

Models for change management are proven frameworks that can guide an organization in navigating its activities during a change—in a process, in the organizational structure, in the current business model, in the organization's culture or in digital transformation. Right from the time the change is contemplated, through the core element of implementing the proposed or desired change, to the point change has been validated and fully absorbed, there will arise quite a few problems, some relatively easy to handle and some quite stubborn. Models have been developed to suit different types and scales of change in organizations of different sizes, some laying emphasis on what to be changed and others stressing on people involved in the change accounting for their feelings and emotions.

One of the earliest models is attributed to Lewin (1950s), which has been recommended for significant changes like organizational re-structuring or work culture shifts. It breaks up the change management process into three steps. First, unfreeze the current system to reveal its strong and weak points and to communicate to the entire workforce 'why' a desired change has to be implemented. This is the step to prepare for the change.

Second, change or initiate and continue to implement the change (which may imply even developing the change in concrete terms so that it can be adopted by the organization). Corporate management should continue to communicate and provide support to employees for effective implementation of the change. In fact, this step or, rather, phase is generally prolonged to accommodate and resolve different sorts of problems at individual and team levels about the content of the change and the best way to implement the same.

Finally, freeze the changed system, so that it does not roll back to its earlier position, by rewarding people involved in change implementation and absorption, taking appropriate safeguards through documentation, training, resource provisioning and supervision and control, deployment of key performance indicators and continuous feedback to employees.

McKinsey's 7S model is meant for large organizations undergoing strategic transformation like merger or acquisition. It addresses seven elements: (i) strategy (what the organization wants to achieves and how it plans to achieve), (ii) structure (command and control mechanism, reporting system, allocation of jobs and responsibilities among employees), (iii) systems (formal processes carried out and tools deployed like technology, performance evaluation procedure, budgeting procedure), (iv) shared values (core beliefs and principles reflected in the organization's policy and strategy and imbibed by leadership), (v) skills (including ability to think differently and out-of-the-box), (vi) style (leadership behaviour and management approach) and (vii) staff (strength, alignment with organizational values, commitment and engagement). The first three are 'hard' elements, which normally are influenced by top management (initiating the change) and hence are easier to manage. The remaining four 'soft' elements pose problems.

Among several models which are people-centric and take into account the psychology of the people involved in the change process and its outcome, Kotter's eight-step top-down model has been well absorbed by large organizations undergoing significant change like digital transformation or cultural overhauls. This takes a top-down approach in the sense that it is focused on leadership and does not highlight the role of employees, especially in terms of their involvement initially to comprehend and vet the proposed change. It goes through eight stages as stated below:

1. Create a sense of urgency (for the proposed change) in the entire organization.
2. Build the change team, preferably cross-functional and multilevel, with members sold to the idea.
3. Form a strategic vision about the change process and its outcome.
4. Communicate the vision among right stakeholders, stressing on benefits to accrue.

5. Remove barriers to change through presentations harping on how everyone benefits, directly and indirectly.
6. Focus on short-term win, reward people involved, and prod for effective long-term sustenance.
7. Maintain momentum by supporting employees and providing resources needed and
8. Institute improvements (may be incremental) to make the change more effective in the long run.

One point about this model is the absence of providing feedback to employees and benefits assured in the beginning. This may attract employee resistance or at least lack of commitment to sustain the change.

The nudge theory, as stated earlier, has worked quite successfully in organizations across the spectrum, especially in the case of health or safety or sustainability initiatives. The significant elements in this model are *define* the change that you like to implement and think about it from the employees' point of view. *Present* the 'change '(what to change and why) based on the likely benefits to employees, in terms of financial gains and/or by way of job simplification and reduced physical labour. *Treat* the change as a choice recommended logically and not as a command imposed forcibly. *Listen* to employees' feedback throughout the process and even after it has been incorporated.

Bridge's transition model is a feelings-based personal approach in which the organization has to help employees through three phases: (1) ending, losing and letting go—all reflecting insecurity from change; (2) the neutral zone between welcoming the new and sticking to the old; and finally (3) the new beginning. This approach is also suited to the case of significant changes likely to take employees by 'fear' of losing position, prospect and even the job.

The change resistance model spells the chaos likely to prevail initially through lack of coordination among leaders and members not all fully aligned with the purpose of the change being introduced and its likely impacts. As chaos cool down and the integration of the desired change goes through, eventually the changed situation becomes the 'new status quo'. The focus in this model lies in pre-empting or at least tactfully and quickly handling 'chaos' through adequate communication, training and guidance, supervision and support.

Some more models have been proposed on more or less similar lines. Deming's P-D-C-A (Pla, Do, Check, Act) model has also been associated with change management, not without any reason. This model has been described in some detail in Chapter 8, dealing with process management.

11.6 TOC AND CHANGE MANAGEMENT

The theory of constraints (ToC), propounded by Goldratt and discussed and debated by several scholars and professionals, has been accepted as a management tool (complete in terms of concepts, definitions, measures, methods and tools) to improve organizational performance under various types of constraints. It makes use of an effective and advanced problem-solving methodology known as the thinking process. Improvement necessarily implies change, and change in the organization as a whole, in a selected activity or even in strategies or behaviours. The problem of managing such a change to respond effectively to problems of constrained performance—especially in the realm of processes and projects with the associated impacts—without getting bogged down to efforts to overcome resistance has been studied in ToC. Thinking process is a generic approach used for investigating, analysing, and solving complex problems (Rahman 1998). In essence, they are 'cause and effect' tools, which help in identifying core problems or constraints, finding win–win solution and implementing them to eliminate the problems without generating any new constraint. These tools enable the management to analyse complex systems having multiple many interdependencies by addressing the following three key issues:

- What needs to be changed?
- What should it be changed to?
- What actions will cause the change?

ToC makes use of a sequence of simple 'effect-cause-effect' (ECE) diagrams based on relevant logic, with components called current reality tree, future reality tree, prerequisite tree and transition tree, along with the ECE audit process, known as the conflict resolution diagram (Spencer and Cox 1995). The effect–cause–effect method validates a cause–effect relationship by distinguishing the core problems. It helps in pinpointing the core problems even when the existence of the proposed cause is not directly observable or constraints are temporary in nature. This is done by relying on the intrinsic logic of the situation itself for pinpointing the constraint. It begins with an undesirable effect, speculating a plausible cause and validating this cause by verifying another effect which results from the same hypothesized cause. The existence of a second effect could only be there if the speculated cause actually exists (Sullivan et al. 2007). The process of hypothesizing a cause for an undesirable effect and then finding another effect arising from the same cause is known as 'effect–cause–effect' phenomenon (Goldratt, 1990), as shown in Figure 11.1

ECE diagramming is a powerful way of uncovering root causes for an undesirable effect in any organizational setting. In fact, it is said to be the only method which can effectively identify the constraints like policy or paradigm constraint, which do not result in permanent constraints. The details of logic tools used in TP and their purpose are described in a snap-shot manner in the Table 11.1.

In this regard, a set of rules were also developed to improve communication and verify logical arguments of different people while applying the thinking process. These logic rules are called the categories of legitimate reservations (CLR), which provides a well-defined methodology for finding errors in the logical reasoning. It relates to entities or statements in a logic trees and diagram built in thinking process. A legitimate reservation exists if the logic that is used does not make sense (Rahman,

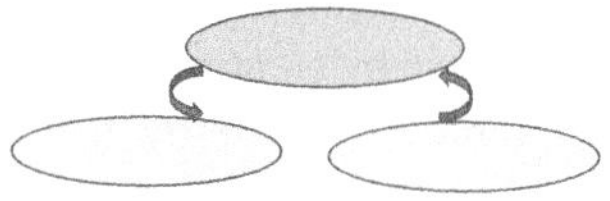

FIGURE 11.1 Effect–cause–effect diagram.

TABLE 11.1
Diagrams used in ToC

TP Tools	Purpose	Role
Current reality tree	Decide what to change?	Pinpoint core problems
Conflict resolution diagram and future reality tree	Decide What to change to	Create solutions
Negative branch reservation	Decide how to change	Audit/verify the solution
Prerequisite tree, transition tree and strategy and tactic tree	Decide how to change	Implement the solution

2002), for example, the cause or the effect does not really exist or the cause–effect relation is clear or some relation connecting a cause with an undesirable effect is misplaced or the predicted future effect of changing a cause may not be unique as hypothesized, and the likely reality tree: Current reality tree (CRT) captures the present state, that is, the state of realty as it exists today. An existing condition is called reality (Goldratt, 1990). Thus, it helps in uncovering what is unsatisfactory and needs to be improved. While developing the CRT diagram, symptoms of the problem characterizing the undesirable state or situation are identified first and then traced back to their root cause by adopting a top-down approach. The most probable chain of cause and effect are identified for the set of existing circumstances. The relation between undesirable effects and postulated likely causes are tested by applying CLR or using effect–cause–effect logic. It usually includes one feedback loop, which will form vicious circle leading to Undesirable Effect (UDE) (Mabin et al., 2001). It can assist in pinpointing the policies, measurements and behaviours, which contribute to the existing UDEs (Burton-Houle, 2000). In addition, it can also help in (a) developing the understanding of complex system, (b) identifying UDEs, (c) pinpointing the core problems, (d) identifying the constraints that must be dealt with first to maximize the benefits and (e) picking up the one simple change that can bring about greatest improvement.

Conflict resolution diagram: Conflict resolution diagram enables organizations to resolve the conflicts which perpetuate undesirable effects. It is also known as the evaporating cloud method. This is a necessity-logic-based tool forms an intermediate stage of thinking process while progressing from CRD to future realty tree. Conflicts usually arise from false assumptions, beliefs or myths, which unnecessarily constrain project work. Here, the analyst resolves the conflict by first exposing the arguments of the two sides and then by adding injections, that is, solution, or remedy. Injections can be defined as actions or conditions which are not present now, but they are necessary to attain a desirable future condition or goal. This logic tool can help in identifying the conflict, explaining the problem, identifying assumptions creating the conflict, resolving the conflict, avoiding the compromise by creating win–win situation and finding breakthrough solutions. CRD and prerequisite tree are built by using the necessity-based logic. The necessity-based logic implies that 'in order to achieve, X one need Y'.

Future reality tree: It depicts the desired future states or could-be future. It also graphically describes the results of injecting changes that have been designed to eliminate UDEs. Therefore, it provides answers to the question, 'change to what?'. It may reflect how undesirable effects (UDEs) have been turned into desirable effects (DEs) by undertaking injections. A future reality tree (FRT) structure consists of a cause and an injection linking to a desirable effect. In essence, it serves following important objectives and (Mabin et al. 2001):

a. Testing the effectiveness of new ideas before implementing them
b. Serves an initial planning tool and find whether proposed changes will yield desired effects
c. Finding whether proposed changes will create new or collateral problems
d. Assist in making desired effects self-sustaining by using positive reinforcing loops
e. Help in assessing the impacts of localized decisions for the whole system
f. Persuading and convincing management to support a desired course of action.

The current and future reality trees and the transition tree are developed by using sufficiency-based logic (cause and effect). Sufficiency-based logic can be of following three types:

a. 'X is sufficient to cause Z'.
b. 'If both X and Y occur together, they will be sufficient to cause Z.'
c. 'X and Y (separately) both contribute to Z, and between them are sufficient to cause Z.'

Prerequisite tree: The prerequisite tree (PRT) is generally used to analyse the complex systems and identify the obstacles that might prevent the injections from being implemented. It is a necessity-based logic tool which helps in finding the required remedies to neutralize these obstacles. The PRT starts from an intermediate objective, which might be an injection that has been developed from the FRT. The intermediate objectives can be considered as the sequential steps designed for implementing the change or achieving the final goal. It is worth mentioning that the obstacle may be a real constraint or a resistance to change which has been planned.

Negative branch reservation: A negative branch reservation is said to be a modification of the future reality tree. However, it really deserves to be treated separately as it brings out new negative outcomes that might result from our injections. The negative branches will uncover both obvious and unanticipated outcome. The obvious one will be verbalized as a reservation, which have been given 'yes' at some earlier stage, 'but' now has to be dealt with the reservation.

The unanticipated one should also be removed or mitigated by using this tool in a systematic way. The potential UDE can be either trimmed or modified by amending the injection or adding another injection. Put simply, this will prevent a situation where a new issue would crop up while implementing a countermeasure to an identified constraint. Thus, negative branch reservation can be considered a robustness test usually associated with proposed injection. A negative branch reservation diagram is shown in Figure 11.2, which shows that injection 3 intended to give desirable effect leads to an undesirable effect. It indicates that negative branch though unanticipated

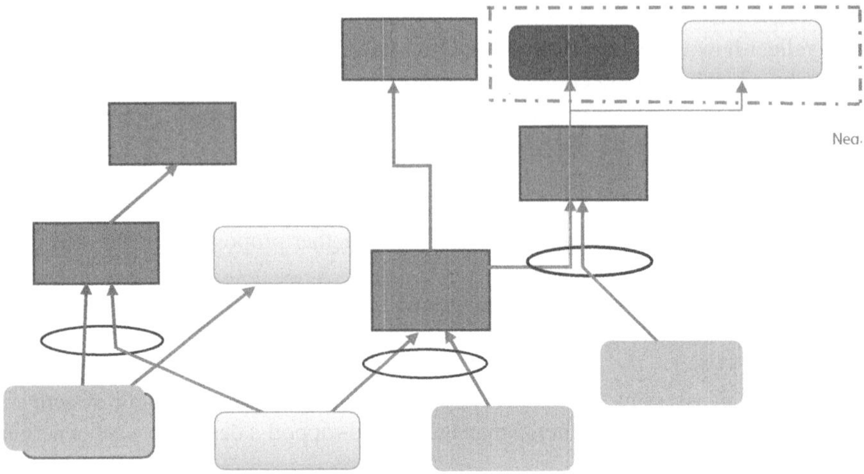

FIGURE 11.2 Negative branch reservation diagram.

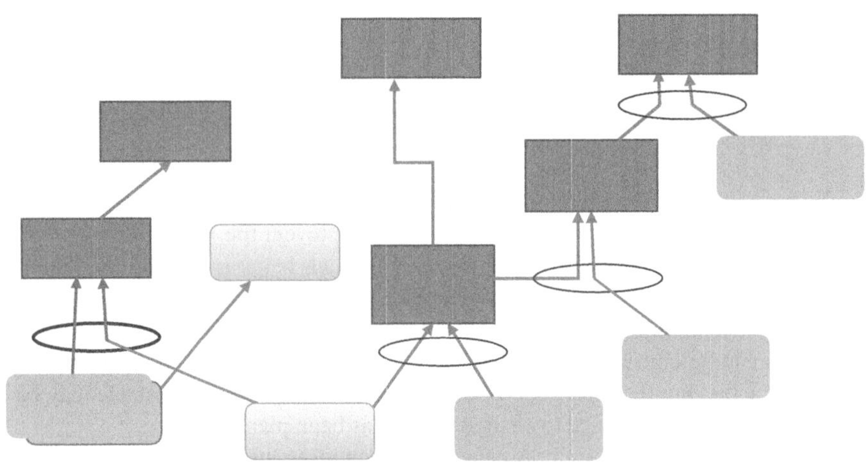

FIGURE 11.3 Trimming of negative branch.

has 'developed' as an interaction of injection 3 and neutral effect 3. As a result, this negative branch is required to be trimmed by an additional injection or a modification of the original injection prior to the negative effect. An example of trimming is shown in Figure 11.3 where injection 4 is used to take care of an undesirable effect. The ellipse is used to represent a logical AND connection with sufficiency condition of causality relation by embracing two or more arrows connected to an entity.

A rectangle or rectangle with rounded corners graphically describes an entity, that is, part of the system which is being studied or an action in the TP diagram.

Strategy and tactics tree: The strategy and tactics tree actually depicts the pathways for achieving the desired goal. It provides a logical architecture which efficiently organizes knowledge and plans to bring in proposed changes. Goldratt has defined strategy and tactic and these definitions, though seemingly simple, are found quite useful in solving problems and managing project progress. 'Strategy' can be considered as the response to the query 'What for?', that is, the objective of a desired change whereas "tactic" provides pathway indicating, 'How to change to the desired state?'. Thus, tactic should be outlined for each strategy while developing this tree. The strategy and tactics should exist as 'pairs' for various levels of the project activities. The strategy and tactics tree therefore outline the detailed hierarchy of the proposed changes along with implementation sequence and their interrelationship that should be followed to successfully traverse the trajectory of changes. S&T tree will bring out and eliminate the conflicts which might be arising from the misalignment of activities with organizational goals and objectives.

11.7 COMPLEXITY OF CHANGE MANAGEMENT

Change is an extremely nuanced phenomenon and change management is an extremely complex task. This is evident from the fact that transformational changes and even other forms of planned change have failed to achieve the expected results in a considerable proportion of cases. An indirect but interesting reflection of this complexity is in the diversity of discussion on the subject—from theory of disequilibrium and punctuated equilibrium in physics to the roles of leadership and organizational resilience in management. Organizations—especially those in the service sector—are quite often complex, adaptive and intelligent systems which operate in networks—local and also global—of providers of inputs to their activities and recipients of outputs. Sometimes peers as well as competitors may also join the networks. Change initiation and management in such systems is fraught with many problems. The usual 'command-and-control' leadership does not deliver the desired results. Even strict adherence to a prescribed sequence of steps does not produce direct linear outcomes. Self-organization and social co-ordination within the organization through peer influence preceded by the alignment of a few committed protagonists becomes a necessity. Careful feedback of results flowing in to hold on the motivation for change must be ensured and creativity has to be promoted. Complexity theory does not provide for a central control, though some form of guidance and supervision cannot be ruled out in managing organizational change.

Surprisingly, seven has been a common figure in dealing with change management like the seven myths about change management mentioned by Jarrett (2003), seven R's of change management or the seven C's in this context. Of course, this is just an illustration of verbose formulation of some simple ideas and thoughts characteristic of many publications in management literature. Managing a planned change to bolster growth or march towards excellence faster than earlier through a 'big' organization-wide change is more problematic than a change which has to be initiated for the sake of survival in the context of some changes that have shaken the business environment and to which resistance is likely to be much less.

The present author is tempted to recall the seven myths about change management as quite a learning process. Not to distort the original presentation by Jarrett (2003), the myths are quoted verbatim with some brief explanations provided by the author.

Myth 1: 'Organizational change management creates value'. Quite true if 'change' has been properly understood and change management remains aligned to that understanding, along with an appreciation of the concomitant risk. Else, value is likely to be lost.

Myth 2: 'Resistance can be overcome'. Resistance is deeply entrenched in every aspect of the organizational structure and individuals have a general tendency to resist change, especially when such a change has not been planned with their involvement and the change desired, prompted or imposed by the top management seemingly affects status quo in regard to work practices. As argued by Jarrett, change should be planned when an evaluation of (a) dissatisfaction with status quo in terms of organizational performance as indicated by some relevant metrics, (b) expected benefits from the vision of proposed change and (c) agreed processes that remove obstacles blocking access to the desired state exceeds the cost of change (considering direct as well as indirect costs). To the first three factors, 'resistance' has to be added as a deterrent to access to the desired state as envisioned in the proposed change. To overcome 'resistance', a lot of effort has to be put in and that means cost. Such an exercise is rarely done in its entirety, since the evaluation is quite complex.

Myth 3: 'Change is constant.' Considering both transformational or breakthrough and incremental or sustaining changes as constituting a 'punctuated equilibrium', change should remain a constant feature of any organization.

Myth 4: 'Change can be managed.' Given the complexity of change, it may be taken as a wise advice that change cannot be possibly 'managed', partly because outcomes as expected may be divergent and even unexpected. However, proponents of change can stimulate the change and steer it through the anticipated 'rough' weather.

Myth 5: 'The change agent knows best.' To be remembered is the fact that even the change agent is himself or herself constrained by his or her inherent resistance to the proposed change. Of course, if the agent is the person or group who has deliberated duly to propose or promote the change, we may reasonably expect that the change agent can be approached for necessary guidance in implementing the change.

Myth 6: 'Accepted wisdom is to follow the steps.' As is widely known, change is an iterative and dynamic process and such a process should be carried out in steps following a proper sequence. There exist several such step sequences suggested by exponents of change management or a sequence of steps best suited to the change to be absorbed can be drawn up and followed. The pointer is the fact that no prescribed sequence of steps may fit a specific type of 'change' in a specific organization with its own culture and structure.

Myth 7: 'Big changes require big changes.' It must be borne in mind that even small or incremental changes can deliver significant results. And small changes are easier managed, in the face of weaker resistance and in the context that often such small changes are suggested and brought about by shop-floor level people themselves.

The list of 'myths' named by Jarrett are essentially cautionary advisories to proponents and leaders of change so that they can prepare themselves better in terms of the timing to announce a change, expected results to be moderated and duly communicated to the workforce to win over their conviction, confidence and support. The list also bestows on the change agents the responsibility to carry out adequate fact-based analysis of the current as well as predicted future situation within the organization and the environment outside. Resources—material, technological and human—required to implement the change have to be carefully estimated and duly provided to the team. Resilience has to be built up to overcome any unlikely or unexpected outcomes of change.

The seven C's or the seven R's are requirements to manage a planned change and have been already discussed in earlier sections in this chapter.

12 Ethics and Values, beyond Technology

12.1 ORGANIZATIONAL CULTURE

Beyond technology and tools, intellect and intuition lie the pivotal role of organizational culture in the pursuit of excellence. In fact, organizational culture and organizational climate are two dominant determinants of organizational performance. Climate and culture are two distinct but inter-related constructs. Culture, inherited from founders, architects or promoters, moulds the vision and mission of the organization and therefore influences the policy on an enduring basis. Culture also permeates in terms of perceptions by employees across the organizations about current work environment practices. Since the latter directly affect performance on the tasks, operational performance largely depends the organizational climate. A carefully crafted culture along with a conducive climate is absolutely needed for any organization aspiring to achieve excellence in future.

Culture is a complex and potentially challenging construct. It has been variously defined by different thought leaders in the field and practiced differently in different organizations. However, its role in influencing organizational performance right from strategy formulation to planning at different levels and implementation has been acknowledged by all the concerned experts and practitioners. One of the most convincing writings on the topic, according to the present author, is the one by Schein (1984). He offers a formal definition of organizational culture as

> the pattern of basic assumptions that a given group has invented, discovered or developed in learning to cope with its problems of internal integration and external adaptation, that have worked well enough to be considered valid, to be taught to new members as the correct way to perceive, think and feel in relation to those problems.

Schein amplifies the meaning and scope of each minute component of this longish definition. Starting with a dichotomy of values as ultimate, non-debatable, taken-for-granted values which are unconsciously practiced against debatable, overt and espoused values, he refers to the former set of overt values as 'assumptions' and refers to the second set as 'values'. While we take up values for a discussionsubsequently, right here we note that organizational culture should be based on a value system

DOI: 10.1201/9781003591733-12

accepted with a sense of commitment by organizational leadership to adhere to the norms and values embodied in a charter or similar document.

The culture of an organization is visible in its architecture, technology, layout of production and office units, patterns of dress, visible and audible behaviour patterns and through public documents like charter, employee orientation manuals, stories, etc. and revealed in events organized to focus on out-of-work life to make life more enjoyable. This aspect of culture is easier to comprehend, compared to an understanding of its value system. Cultural paradigms differ across regions, from one type of control over business by the state to another, and over different work practices. The most important fact to keep in mind is that even the strategy adopted by an organization to ensure a smooth journey in future has to take due account of organizational culture. It is the culture that determines success or failure of a strategy and hence the strategy cannot be oblivious to culture or ignore it as a removable burden of legacy.

The policy and strategy adopted by an organization reflects the organizational value system and organizational ethics. These values and ethics could be documented or could be practiced without being written down and disseminated to all within the organization. It has to be admitted that an individual working in an organization may have a value system or may follow a code of conduct that differs from the organizational value system or code of conduct, and this may affect the individual's actions and the organizational performance. This implies an important role of leadership to align individual values with organizational values through logical persuasion and demonstration of possible consequences of alignment (at least within work area) and non-alignment.

The rising phenomenon in which some organizations encourage a value system that may not respect age-old virtues like honesty and adopt unethical practices to demonstrate high levels of achievement stresses the role of values and ethics in determining or influencing organizational performance. These values and ethics (or the absence thereof) are reflected in the organization's policies and strategies.

When the leadership is convinced of the role of transparency, ethical conduct and basic values to guide the entire organization and all its activities, and also mandates the workforce to adopt organizational values, including basic or core values and ethical conduct, practices are found to vary across organizations, even within the size, class or with the same product and service profiles. A value may be viewed as a conception of what an individual or organization regards as desirable and in terms of which selection of the means and ends of action is made from the available alternative modes.

Values can be classified as absolute and instrumental. Absolute values like honesty and integrity, modesty or humility (but not self-negation), compassion are accepted—rather respected—in all ages and across human societies. However, values like diligence, perseverance, courageousness, passion, self-confidence and the like are instrumental values which tend to help an individual realize their goals and objectives. Transparency and consistency in behaviour and action definitely qualify as values and there is no imperative need to classify all values as either absolute or instrumental.

Incidentally, the author likes to recall a personal experience and a statement made by an acquaintance quite some years back—with regard to the promotion of values by leaders in some business organizations. The second one relates to a group of business leaders discussing the role of honesty as influencing business policy in a roundtable event. Members, except one, agreed that 'honesty is the best policy'. The lone dissenter remarked, 'No. Honesty is the only policy'. The first incident took place following the purchase by the author of a simple calculator for his schoolgoing son when scientific calculators had not entered the market. It was an inexpensive item bought from a small stationery shop in a foreign country. On return, it was found that one side key was not operating, maybe due to mishandling during transit. It so happened that te author had invited a scientist from the country where he purchased the calculator to a conference organized by the author. During an informal chat, the author mentioned about the defect in the calculator and the scientist friend immediately wanted the item from the author so that he could organize a replacement. Taken aback by the offer, the author pleaded that there was no need to get a replacement since the item was not at all an expensive one. The friend argued that it would poorly reflect on the trade behaviour of the seller if this incident was not brought to the latter's notice and proper action. Strangely beyond the author's expectation, a replacement was sent to highlight the national character of the trading community in that country. In both these two cases, Japan was the country and Japanese were the people in the roundtable. The author hastens to add that this recollection is just a stray remembrance not at all meant for any generalization.

Corporate social responsibility (CSR)—as a concept and as an obligation in recent times—reflects on the ethical requirement of any organization, especially for-profit business organizations, to acknowledge the contribution from society (taken in a broad sense) as providing the biggest support to an organization and consequently to contribute to overall development of society for sustained growth and development of all concerned. CSR has been directly or in a covert manner included in all current versions of organizational excellence models in the 'Results' part. However, the present author strongly feels that CSR should be considered as an important enabler of organizational excellence, and not just a component of results.

12.2 ORGANIZATIONAL CLIMATE

Organizational climate has been defined as the totality of shared perceptions of employees about the work environment and how it affects their well-being. Thus, it is a set of attitudes and experiences which label the organization in terms of both static characteristics (a degree of autonomy for example) and behaviour outcomes. It is a set of specific attributes of an organization that may be induced from the mode in which the organization deals with its members and with its environment. Rules (official or otherwise), procedures, practices and everyday habits (inside the workplace) reflect on the prevailing climate.

Organizational climate can and should be positive to enhance people engagement and improve performance. In a positive climate, employees feel cared for, happy and motivated while in a negative climate they feel stressed out, discontent and detached. Beyond the impact on employees, the factors which characterize the prevailing climate deserve attention of concerned leadership. In this sense, one can recognize the following types of climate prevailing in different organizations pursuing different cultures and guided by different types of leaders.

- *People-oriented climate*: Allows flexibility and freedom to employees within specified boundaries, taking care to ensure the preservation of order and discipline and promoting creative ideas
- *Rule-oriented climate*: Enforcing rules strictly to ensure adherence to norms and organizational values
- *Innovation-oriented climate*: Promoting, supporting and rewarding ideas, thoughts and concrete suggestions for improvements in the existing processes and products or services and for creating new or novel products and service. In this climate, people are freed from routine, repetitive tasks as much as possible by adopting automation and related digital technology and are encouraged to think about the existing processes and their outputs and possible improvements thereon
- *Result-oriented climate*: Focusing on the achievement of targets related to outputs in terms of speed (rate of production), quantity and quality. High-performing individuals and teams are rewarded and underperforming ones are either supported to improve their future performance in terms of deputation to a training and skill development programme or are warned of consequent actions including transfer to some other jobs.

It should be remembered that the above categories are not mutually exclusive nor are these exhaustive. Thus, a people-oriented climate can well be result-oriented also and similarly a rule-oriented climate may also be result-oriented. Only priorities differ and possibly an innovation-oriented climate cannot necessarily be result-oriented. We can characterize organizational climate from a different perspective as

Networked in terms of necessary and desirable interactions among groups or functions by way of communication, consultation, co-operation and collaboration, primarily to improve performance of each of the groups or teams or even individuals in their respective areas.

Fragmented by way of barriers put up between individuals and groups, allowing interactions only when necessary and duly authorized. This helps leadership to fix responsibilities for any lapses and, hopefully, to identify and appreciate successful ones.

Communal in the sense that each employee has a strong sense of belonging to the organization, looked upon as a well-knit community. (The adjective 'communal' used by some authors may not be an appropriate one.) This is quite the desired climate.

Mercenary is the word used to characterize an organization that cares only for financial performance as of now, without adhering to principles of propriety and recognition of people's contribution. This adjective is not a logical conclusion based on evidence and experiences, but more like an exclusion from other categories.

Factors affecting the organizational climate which can be modulated by top leadership to create and sustain a congenial climate include the following:

- Organizational context: revisiting the vision and mission as accepted initially as well as the goals and objectives considered in the strategy being currently implemented

- Structure of the organization indicating the flow of command and control, the bridges and barriers across functional areas and teams.
- Leadership style and processes to guide, support and control operational teams and aligning them with the organizational values
- Communication patterns to keep employees duly and correctly informed about any changes being initiated—big or small—in which they will be involved, besides communication among employees and between a member and the leader on work-related or even on social issues
- Work–life balance in terms of duration of work and work-related stress telling upon employees' attention to and family members and social engagements
- Physical environment with emphasis on occupational health and safety measures, ergonomically designed workplace in terms space for movement, adequate illumination and ventilation, conveniently installed equipment, adequate conveniences and control over dust and noise pollution, emergent first-aid arrangements, etc.
- Diversity, equity and inclusion in terms of members belonging to different genders or speaking different languages or practicing different faiths being treated equally and top leadership encouraging and promoting diversity, say, during celebrations and social events.
- Value system and norms as are communicated and explained to all employees with a message to align individual values with organizational values, at least within the ambit of organizational affairs. The manner in which dissent is treated and attempts are made to remove such dissent also plays an important role.
- Technology adopted in various functional areas coupled with exercises to enable people in meaningfully using it is a key factor in realizing the impact of climate on productivity and hence on organizational performance. The manner in which technology reduces work stress and promotes free thinking and facilitates employees to develop themselves affects climate.

Culture and climate are intertwined to influence organizational performance. However, at least for the sake of comprehension and subsequent modulation, it may be worthwhile to note some distinctive features of each. Culture is derived from a set of deep-seated beliefs and ingrained values, while climate is based on perceptions. Culture is enduring and prevails over a long time horizon till a significant change in environment mandates a change in culture or when ownership changes are accompanied by a change in beliefs (may be postulated by experiences) and value systems in general.

12.3 INDIVIDUAL AND ORGANIZATIONAL VALUES

Values are beliefs that a person holds about things and aspects of life. These are guiding principles that mould a person's desires, feelings and actions all his life. (Values are acquired or formed over years of life in different ways.) To one facing a dilemma, these values guide one's choice of action and behaviour. Different societies have different value systems and the behaviour and actions of people in a society are guided by these values.

People also have their own value systems, which act as the lens through which they view the world around them and which provide them a basis to make judgements. This also brings in some complexity in situations where a contradiction arises between values of the society and personal values held by an individual. Such conflicts create problems in the society and the individual.

Organizations also have their respective value systems which are meant to guide individuals in so far as they as they work for the organization in their actions and behaviour within the organization as well as their interactions with the environment within which the organization operates. In this sense, the same organization may pursue different value systems when operating within different environments.

People management activity has the important dimension of resolving any conflict between the values cherished by an individual within the organization and the organizational value system. This is not an easy task, given the problem of bringing out value systems of different persons—not always overt. Resolving any conflict leads to more difficulties. Just to illustrate, the following was accepted to define the core values of a service organization to be deployed among its people at all levels:

1. Customer sovereignty: wWe owe our existence to the customers.
2. People Orientation: People constitute our greatest asset. We have to respect other individuals.
3. Innovation and entrepreneurship: A great idea is bon and becomes a reality where the spirit of entrepreneurship is kindled and innovation fostered.
4. Transparency and integrity: Integrity may be difficult to define, but lack of integrity is better understood.
5. Anticipation, speed and flexibility: Problems lurking inside should be anticipated, no opportunity to act should be delayed and scope for modification has to be built in.
6. Passion for superior performance: To win, one has to possess a fire, which should be ignited to achieve better performance.
7. Social and environmental concern: Every one should contribute to the development of the society and protection of the environment.

An absolute virtue like 'humility'—properly interpreted—may be an important value for a learning organization. Humility is not self-negation but definitely implies respect for others. Quite often, humility may be motivated by a sincere admission of one's limitations (internally, if not in public) and a genuine effort to learn more from its own experiences and from others.

12.4 ROLE OF ETHICS

Ethics correspond to codes of conduct that decide what is wrong and what is right in a particular circumstance or in every situation. Also called morals, these codes have slowly and gradually evolved to become accepted norms and forms of behaviour. In different societies, there are different sets of ethics, though some morals are considered to be universal and are accepted as such in all societies.

In an organization, ethics are unwritten rules that people in an organization should abide by. In this context, ethics tell the people what is right and what is wrong for the organization. In different professions and in different environments, we find different accepted codes of conduct or ethics. Here again some moral codes could be a part of the codes followed by all such bodies.

Unlike value systems, which can ostensibly differ among organizations within the same environment, ethics or codes of conduct remain more or less the same.

While value systems and codes of (ethical) conduct are context-specific, at least for the sake of argument, there are virtues, such as honesty, diligence and perseverance, which stand out as universal and eternal. These are incorporated into different extents in the codes of conduct or value systems by different organizations.

What should matter in the long run is not just performance, but the way this has been achieved. Thus, performance may have to be discounted for any deviations from the declared value system and from the code of conduct that is acceptable to the environment. Discounting not just in monetary terms but accounting for social impacts has to be put in place.

Discounting practices cover a wide spectrum, right from full impunity to severe penalty. This is why a duly penalized performance measure that can be used for comparison across the board is difficult to obtain. But the best bait is to go by ethical practices and value-based actions and reactions is not the ground reality in the contemporary world that believes in bending rules and cutting corners to perform 'better.'

Situations arise where the organization has to use enough discretion linked to its business performance and rights and privileges of workers that are not all unethical. Thus, in a research and development organization involved in carrying out some experiments and analyses, some dedicated workers who design experiments and analyze findings thereof should qualify for intellectual property rights possibly. Even when this activity has been undertaken on behalf of a parent organization which provides funds to and/or guides activities of the organization, the findings cannot be disclosed to anyone else before being communicated to the parent and without the consent of the parent. Thus, while disclosure to the public at large of findings and results generated by an academic institution or an official agency meant to provide such findings and results for consumption by the concerned user group, it will be going against ethics to divulge this information without taking into confidence the rights of individuals and groups behind generation of such information.

12.5 VALUE-BASED LEADERSHIP

Quite a few styles of leadership emerged over the years and were being practiced by leaders with varying degrees of success in various organizations. The story included several cases where organizations had surprising growth and yielded surprisingly improved performance under some forms of leadership such as transformational leadership, which could not be sustained. In fact, the urge to get quick returns at any cost and by any means—not necessarily caring for professional or business ethics and bending rules and regulations—led to quite a few scandals and brought to surface the need for a more scrupulous leadership that would enable organizations to achieve their goals and objectives on a sustained basis. Issues of distrust, lack of morality and workplace incivility surface as threats to higher levels of organizational performance and as pointers to the failure of contemporary leadership practices. The search for new forms of leadership proceeded gradually and one outcome was the unifying theory of value-based leadership. Informative and comprehensive reviews of several aspects of value-based leadership are available in Copeland (2014) and Sumanasiri (2020).

Value-based leadership is a leadership style that builds itself on a set of guiding principles that reflect the core values and beliefs of the organization, laying down clearly what is acceptable as 'good' for the organization and what is not acceptable. At this point, it must be emphasized that nothing unethical can be treated as acceptable, for example professional ethics for different categories of professionals engaged in the organization and business ethics requiring no harmful effects of activities by the organization in the business environment, society and the environment. These core values should be accepted and practiced by both the leader and the team members. It serves as a framework for making decisions and interactions (internally as well as with the environment), shaping the behaviours of the workforce and guiding their actions including those by leaders. Value-based leadership does not focus on personal gains (even when a positive image could be derived), instead tries to focus on achievement of organizational goals and objectives. The theory of value-based leadership draws upon some appealing features of several other styles of leadership such as transformational leadership, authentic leadership, responsible leadership, spiritual stewardship and servant leadership.

To ensure a reasonably good performance on a sustainable basis, true leadership has to be based on a set of values like transparency, integrity, trust, mutual respect, belief in trusteeship, non-ownership of property and empathy. These values must be practiced sincerely. A few successful organizations bear testimony to this fact. Only professing a set of core values read out routinely on some occasions will not do. The basic requirements for value-based leadership practice in any organization, which expectedly has a set of core values and beliefs on the basis of which the organizational policy (at the corporate level) and strategy are clearly spelt out, are that the leaders internalize the core values of the organization and align their personal value systems with that for the organization. Such a practice is revealed through

Continuous communication involving leaders and team members to help members fall in line with the organizational values.

Leaders should demonstrate their absorption of the organizational values through their decisions and actions.

Leaders holding themselves and others accountable for following the organizational values, some of which could be beneficial to employees also. Core values have to be articulated verbally among all and demonstrated in practice by leaders.

Value-based leadership (sometimes mentioned as values-based leadership) calls upon leaders to lead by example in terms of their behaviour, decisions and actions, which demonstrate their allegiance to the values of the organization. Incidentally, values in an organization are not always meant to extract as much as possible from employees, disregarding their health and personal life needs. The author recalls the case of an industry in the telecommunication sector which had a strict norm that one day a week, no employee will be allowed to stay back in the office for any work, however urgent or important. To make this choice a mandate, leadership would swithch off all connections and withdraw all facilities beyond the specified office hours. Leaders must engage team members frequently in discussions on values and ethics to remove any doubts or any issues with the value system of the organization.

All the above will help development of a culture of trust and transparency and create a sense of shared values and vision between the leaders and team members. Such a collaborative and empowering work culture is congenial to creative thinking by employees and take up innovation exercises. As indicated earlier, such a culture is almost sure to generate good and even great performance.

12.6 GOOD GOVERNANCE

Good governance is emphasized least in letters and words in most organizations in both the public and private sector. The principles and practices to ensure good governance are in a way encompassed by value-based leadership, with value system aligned with widely accepted norms and a sincere commitment to that system. A comprehensive understanding of the environment within which any organization operates and the symbiotic relations that can lead to the growth and excellence of any organization is the crux of good governance. Good governance is an indispensable catalyst to push an organization forward in its journey towards excellence.

We speak of governance in the context of human organizations and try to assess different attributes or traits of governance in a particular organization. Some of the attributes are mentioned in the vision or mission statement of an organization, some are reflected in its performance, some get revealed through some internal or external inquiry and others implicitly pervade the organization (like traditions and culture).

An organization is a (social) entity involving multiple persons working towards some common goal(s) interacting with an external environment that affects it and that is affected by it. The environment within which an organization operates is a multi-dimensional concept that combines the concerns, decisions and actions of all stakeholders—from employees or members to investors or subscribers, from regulatory agencies to civil societies, from national or state governments to international watchdogs and court of arbitration.

Any organization affects its environment through the impact—physical, economic and social—of its activities. The environment affects the organization in terms of codes of practice it imposes on the organization.

Organizational governance is the system by which an organization is guided, directed and managed in pursuit of its goal(s) and objectives. Guidance is reflected in developing a vision and a mission statement and sharing the same with all concerned through formal or informal exercises. Direction is contained in policy and strategy, extended further to decision-making principles and procedures.

Management is concerned with the processes of procuring and deploying resources, executing tasks and monitoring and controlling outcomes of various activities. Governance is a ubiquitous concept, applicable to all organizations, irrespective of their nature and the goals they pursue, the size and scale of operation and the environment within which any organization operates. Governance does not just mean managing or administering. The focus of governance is on norms and standards, principles and procedures, as distinct from processes performed or activities carried out. It concerns how these processes activities are planned and how these impact the Environment. Management is characterized by the people who direct and control the processes and who deal with the environment. Management is concerned more with the activities carried out and the outcomes achieved. Deployment of available resources—physical, material, human and technological—and also augmentation of resources as and when needed are within the ambit of management. Management enjoys authority commensurate with the responsibilities it has to shoulder.

The distinctions between governance and management or administration are quite explicit in the case of some organizations, only tenuous and implicit in others. Governance has been generally perceived as a superior function which guides the organizational activities, rather than controlling such activities, to achieve the set goal(s) and objective(s). The latter is primarily a concern of management or administration. Governance in terms of norms and principles is rather generic, while management in terms of planning, executing, monitoring and evaluating is bound to be more specific. Good governance makes an organization committed to sound operating practices based on a set of values like conscience, openness, fairness, professionalism, mutual respect and mutual trust among all constituents and transparency and accountability in building confidence of all its stakeholders, thereby paving the way for its long-term success. It is a culture that guides the people within the organization at all levels, including the board, the management and the workers or members.

12.7 CORPORATE SOCIAL RESPONSIBILITY

We can speak of three different perspectives for a discussion on corporate social responsibility.

The first is provided by the role played by the corporate sector in terms of contributions made by the Indian industry and businesses in the fields of community development, social welfare, protection and enrichment of environment and generally to national development. Some established companies took up a variety of programmes in this direction quite some years back, much before CSR became formally recognized.

The second perspective is provided by the declaration and adoption in the year 2000 of the eight Millennium Development Goals (MDGs) by the World Business Council for Sustainable Development, which were to be achieved by the end of 2015. Several corporate sector units have since aligned their CSR policies and programmes with these MDGs.

Business excellence models and quality awards provide the third perspective. In these models and award criteria, results achieved by an industry or business house do include 'impact on society and impact on environment' besides employee, customer and other stakeholder satisfaction. In the development of these models, it is argued that a business or industry unit howsoever successful it may be at present cannot thrive or even survive without creating a congenial impact on society and environment and thereby discharging its corporate social responsibility.

Given these different perspectives that originated at different times, one should not construe these to be mutually exclusive and can well take the advantage of deriving appropriate benefits and directions from all the three.

CSR is an issue that relates to how a company manages its business processes to produce an overall positive impact on society. Like most of the modern managerial concepts, CSR also encompasses several definitions, varying across companies and countries. The most widely quoted definition is that provided by the World Business Council for Sustainable Development in its publication *Corporate Social Responsibility: Making Good Business Sense* by Richard Holme and Phil Watts (2000). It reads:

> CSR is the continuing commitment by business to behave ethically and contribute to economic development while improving the quality of life of the workforce and their families as well as of the local community and society at large.

A second accepted definition of CSR is "achieving commercial success in ways that honour ethical values, respect people, communities and natural environment." It implies that CSR means addressing the legal ethical, commercial and civil expectations of society and making decisions that fairly balance the claims of all the key stakeholders. In its simplistic terms it is what you do, how you do it and when. CSR is about businesses giving back to society.

It is important to distinguish CSR from charity and good work, for example, in spending money on community projects, institution of endowments and establishment of foundations, or encouraging their employees to volunteer for community work and thereby creating goodwill in the community, which will directly enhance the reputation of the company and strengthen its brand. CSR goes beyond charity and requires that a responsible company take into full account their impact on all the stakeholders and the environment when making decisions.

This requires them to balance the needs of all their stakeholders, such as employees, customers, suppliers, community organizations, subsidiaries and affiliates, joint venture partners, neighbourhoods, investors, local administration and shareholders with their basic need to make profit and reward shareholders adequately.

Under the CSR programme, business houses can institutionalize voluntarism among employees through suitable recognition and incentives. Internal performance evaluation of employees could recognize community work in many forms, for example, teaching in government-run schools, taking part in NGO activities, empowering women, cleaning parks and other public places, planting trees, volunteering in orphanages, protecting the abused and the like. An innovative way to contribute to society is for firms to buy products from millions of artisans who are at the bottom of the pyramid.

In some sense, CSR is a comprehensive set of policies, practices and programmes that are integrated into business operations, supply chains and decision-making processes throughout the world wherever companies do business and includes responsibilities for current and past actions as well future impacts. In its broadest category, CSR typically includes issues related to business ethics, community investments, environment, governance and accountability, human rights, marketplace and workplace, mission, vision and values.

In the backdrop of a growing global pressure on corporate houses to adopt CSR as an integral part of management policies, CSR auditing and reporting is fast becoming mandatory. To demonstrate good business citizenship, firms can report compliance with a number of CSR standards, including

- Accountability's AA 1000 Standard, based on Elkington's triple bottom line reporting
- Global Reporting Initiative's Sustainability Reporting Guidelines
- Social Accountability International's SA 8000 Standard
- The IS0 14000 Environmental Management Standard

However, it is to be accepted that CSR is difficult to quantify and measure. Many companies now produce externally audited annual reports covering sustainable development and CSR issues, but these vary widely in format, style and evaluation methodology.

The benefits of CSR to business houses vary depending on the nature of the enterprise. There is a growing body of literature exhorting business to adopt measures beyond those from my financial considerations, for example, Deming's 14 Points and balanced score card. An enterprise with healthy CSR practices can benefit in terms of recruiting and retaining value-oriented human resources, proper risk management, and brand differentiation in the perception of consumers getting licence to operate.

CSR is the order of the day. Whether corporate houses like it or not, they have to adapt themselves to CSR practices sooner or later, if they have to gain widespread social acceptance and improve their national and international visibility.

The MDGs are the summary of development goals that form a part of United Nation's road map towards implementation of MDGs by the year 2015. The achievement of MDGs has been pledged by 191 member states:

1. Eradicate extreme poverty and hunger
2. Achieve universal primary education
3. Promote gender equality and empower women
4. Reduce child mortality
5. Improve maternal health
6. Combat HIV/AIDS, malaria and other diseases
7. Ensure environmental sustainability
8. Develop a global partnership for development

The importance of MDGs to business lies in the fact that implementation of these goals will help in creating a more prosperous, stable and inclusive economy, which forms the foundation of a thriving business sector. As such, it becomes essential for the private sector to collaborate with governments and others in the society in tackling issues of poverty hunger, environment degradation, diseases and conflicts in the world. The MDG's have paved the way for formulating a revised and possibly more ambitious set of Sustainable Development Goals (SDG's) set to be achieved by 2030.

Many companies worldwide today are aligning their CSR work to the MDGs, and now to the SDG's, not just for compliance, risk management and philanthropy but also as a matter of corporate social opportunity, embedded in innovation, value creation and competitiveness through now products, services and technology and in some cases even transforming their business models to address social and environmental changes.

References and Suggested Reading

Abdullah, M. et al. (2012) Value-based total performance excellence model: A conceptual framework for organizations. *Total Qual. Mgt. Bus. Excel.*, 23 (5 & 6), 557–572.

Abermathy, W. and Clark, K.B. (1985) Managing the winds of creative destruction. *Res. Policy*, 14, 3–22.

Alders, A.F. (2023) *Cultivating Organizational Excellence: A Practitioner's View*. Springer.

Al Tabbaa, O., Gadd, K. and Ankrah, S. (2013) Excellence models in the non-profit context. *Int. J. Qual. Reliab. Mgt.*, 30 (5), 590–612.

Anthony, S.D. Johnson, M.W. and Sinfield, J.V. (2008) Institutionalizing innovation. *MIT Sloan Mgt. Rev.*, January 1.

Antony, J.P. and Bhattacharya, S. (2010) Measuring organizational performance and organizational excellence of SMEs. Part 2. An empirical study on SME's in India. *Measur. Bus. Excel.*, 14, 42–52.

Barnawi, M.B.O. (2022) Organizational excellence models: Failure and success factors of organizational excellence. *Open J. Bus. Mgt.*, 10 (6), 2915–2985.

Barnett, W.P. and Carroll, G.R. (1995) Modelling internal organizational change. *Ann. Rev. Sociol.*, 21, 217–236.

Beycioglu, K. and Kondakci, Y. (2020) Organizational change in schools. *ECNU Rev. Ed.*, 4 (4), 788–807.

Bhagat, L.N. and Kumar, S. (2020) Measurement of organizational excellence in Indian context: An exploratory study. *Int. J. Engg. Technol. Manag.*, 7 (12), 21–32.

Bhaskar, H.L. (2018) Business process re-engineering framework and methodology: A critical study. *Int. J. Serv. Oper. Manag.*, 29 (12), 527.

Bise, J. and Malagueno, R. (2009) The choice of interactive control system under different innovation management modes. *Eur. Account. Rev.*, 18 (2), 371–405.

Bogan, C.P. and English, M.J. (1994) *Benchmarking for Best Practices: Winning through Innovative Adaptation*. McGraw-Hill.

Brander, G. (2023) *Scenario Planning*. Lettersmith.

Brovkin, S. (2024) Success rate of lean initiatives is hardly over 5%. Linkedin post by Sergei Brovkin.

Buist, W. (2024, March 28) Competition versus collaboration (blog). www.williambuist.com

Burke, W.W. (2017) *Organizational Change: Theory and Practice.* Sage.

Burke, W.W. and Litwin, G.H. (2017) A causal model of organizational performance and change. *J. Manag.*, 18 (3), 523–545.

By, R.T. (2005) Organizational change management: A critical review. *J. Change Manag.*, 5 (4), 369–380.

Cameron, E. and Green, M. (2009) *Making Sense of Change Management*. 2nd ed. Kegan Paul.

Clarke, T.N. (1968) Institutionalization of innovation in higher education: Four models. *Adm. Sci. Q.* 13 (1), 1–25.

Claudia, P., Frey, D. et al. (2009) Leading and managing organizational change initiatives. *Manag. Rev.*, 20 (2), 158–175.

Coeckelbergh, M. (2020) *Artificial Intelligence Ethics*. MIT Press.

Copeland, M.K. (2014) The emerging significance of values based leadership: Literature review. *Int. J. Leadership Stud.*, 8 (2) 105–135.

CORBA Team (1995) *The Responsive Organization: Re-engineering New Patterns of Work.* Policy Publications.

Coulson-Thomas, C. (1997) *The Future of the Organization Achieving Excellence through Business Transformation*. Page.

Covey, S.R. (2006) Servant leadership: Use your voice to serve others. *Leadership Excel.*, 23 (2), 5–6.

Dahlgaard, J.J. and Dahlgaard, S. (2013) Business excellence models, limitations, reflections and further developments. *Total Qual. Manag. Bus. Excel.*, 24 (5 & 6), 519–538.

Davenport, T. (1973) *Process Innovation: Re-engineering Through Information Technology.* Harvard University Press.

Davenport, T.H. (2023, March) How AI is helping companies redesign a process. *Harvard Bus. Rev.* www.hbr.org

Davenport, T.H. and Harris, J.G. (2017) *Competing on Analytics.* Harvard Business School Press.

DeBono, E. (1992) *Serious Creativity: Using the Power of Lateral Thinking to Create New Ideas.* Harper Collins.

Dervitsiotis, K.N. (2007) On becoming adaptive: The new imperative for survival and success in the 21st century. *Total Qual. Manag. Bus. Excel.*, 18 (1–2), 21–38.

Domingo, R.T. (2013) *Lean Management Principles.* Asian Institute of Management. www.rdtonline.com

Duggan, K.J. (2013) *Design for Operational Excellence: A Breakthrough Strategy for Business Growth.* McGraw Hill.

Duggan, K.J. (2016) *Beyond the Lean Office: From Lean Tools to Operational Excellence.* Productivity Press.

Elkington, J. (1999) *Cannibals with Forks: The Triple Bottom Line of 21st Century Business.* Oxford Capstone.

Felicio, J.A., Rodrigues, R., Patino-Alonso, C. and Felicia, T. (2022) Allostasis and organizational excellence. *J. Bus. Res.*, 140, 107–114.

Fell, N. (2007, January 21) Triple bottom line approach growing in non-profit sector. *Cause Planet.* www.causeplanet.org

Frost, J. (2014) Values based ledership. *Ind. Commer. Train.*, 44 (3), 124–129.

Ghobadian, A., Mole, K. and O'Regan, N. (2003) New process technologies: Barriers to adoption in small firms. *Int. J. Manuf. Technol. Manag.*, 5 (5–6), 549–562.

Goleman, D. and Charmis, C. (2024) *Optimal: How to Sustain Personal and Organizational Excellence Everyday* [eBook]. Whiteleaf Publishing.

Greenleaf, R.K. (1977) *Servant Leadership: A Journey into the Nature of Legitimate Power and Greatness.* Paulist Press.

Guerrieri, G. (1965) Some characteristic properties of the exponential distribution. *G. Econ. Ann. Econ.*, 24, 427–437.

Guha, S., Kettinger, W.J. and Segars, A.H. (1996) Business process re-engineering: Building a comprehensive methodology. *Inform. Syst. Manag.*, 10 (3), 13–22.

Halders, A.F. (2023) *Cultivating Organizational Excellence: A Practitioner's View.* Springer.

Hamid, M.R.B.A. (2015) Value-based performance excellence for higher educational institutions. *Qual. Quant.*, 49, 1919–1944.

Harrington, H.J. (1991) *Business Process Improvement, the Breakthrough Strategy for Total Quality, Productivity and Competitiveness.* McGraw Hill.

Hiatt, J.M. (2006). *ADKAR: A Model for Change in Business, Government and Our Community.* Prosci Learning Center Publications.

Holme, R. and Watts, P. (2000) *Corporate Social Responsibility: Making Good Business Sense.* World Business Council for Sustainable Development.

Homel, G., Doz, Y. and Prahlad, C.K. (1989) Collaborate with your competitors and win. *Harvard Bus. Rev.* (Jan–Feb). www.hbr.org

Hubbart, J.A. (2023) Organizational change: The challenge of change aversion. *Adm. Sci.*, 13 (7), 162.

Hultman, K. (2017) Unleashing the power of values. *Org. Dev. J.*, 35 (2), 17–32.

Hultman, K. and Gellerman, B. (2002) *Balancing Individual and Organizational Values: Walking the Tight Rope to Success*. John Wiley.

Jobin, A., Ienca, M. and Vayena, E. (2019) Artificial intelligence: The global landscape of ethical guidelines. *Health Ethics Policy Lab ETH Zurich*. Arxiv.org Preprint Version, https://arxiv.org/pdf/1906.11668.pdf

Joseph, E.E. and Winston, B.E. (2005) A correlation of servant leadership, leader trust and organizational trust. *Leadership Org. Dev. J.*, 26 (1), 6–22.

Kelley, D.M. and Kelley, T. (2014) *Creative Confidence: Unleashing the Creative Potential in Us All*. William Collins.

Kettinger, W.J. and Segars, A.H. (1996) Business process re-engineering: Building a comprehensive methodology. *Inform. Syst. Manag.*, 10 (3), 13–22.

Kluemper, D.H. (2008) Trait emotional intelligence: The impact of core-self evaluations and social desirability. *Pers. Individ. Differ.*, 44, 1402–1412.

Konno, N., Nonaka, I. and Ogilvy, J. (2014) Scenario planning: The basics. *World Fut.*, 70, 28–43.

Kotter, J.P. (2012) *Leading Change*. Harvard Business School Press.

Lai, C.D., Moore, T. and Xie, M. (1998). The Beta-integrated model. In Proc. Int Workshop on Reliability Modeling and Analysis—from Theory to Practice. pp 153–159.

Lasrado, F. (2018) *Achieving Organizational Excellence: F & M Approach for Culturally Diverse Organizations*. Springer.

Lee, R.G. and dale, B.G. (1998) Business process management: A review and evaluation. *Bus. Process Manag. J.*, 4 (3), 214–225.

Leme, G. (2024, July 12 blog) How AI is transforming business process management. www.pipefy.com

Lewin, K. (1947) *Field Theory in Social Science*. Harper and Row.

Macintosh, R. (2003) BPR: Alive and well in the public sector. *Int. J. Oper. Prod. Mgt.*, 23 (3), 327–344.

Mainville, D. (2024, May 27) The use of Artificial Intelligence in business process management (NAVVIA. Com). Linkedin.com

Maranco, R.M. (2024) *Data Excellence: How Data Leadership Can Drive Stronger Organizational Performance*. Kogan Page.

Mauli, R.S., Tranfield, T.R. and Mauli, W. (2003). Factors affecting the maturity of BPR programme. *Int. Jour. Oper. Prod. Manag.*, 23 (6), 596–624.

McGee-Cooper, A. and Trammell, D. (2002) From hero-as-leader to servant-as-leader. In L.C. Sears (ed.) *Focus on Leadership: Servant Leadership for the 21st Century*. Wiley.

Mowery, D.C. and Rosenberg, N. (2012) *The Institutionalization of Innovation, 1900–1990*. Cambridge University Press.

Mukherjee, S.P. and Pal, M. (2007) Distribution of quality-adjusted life of a product that can exist in three states. *IAPQR Trans.*, 32 (2).

Nenadal, J., Vykydal, D. and Waloszek, D. (2018) Organizational excellence: Approaches, models and their use at Czech organizations. *Qual. Innov. Prosper.*, 22 (2), 47–64.

Nieto, M. (2023) From R&D management to knowledge management: An overview of studies of innovation management. *Technol. Forecast. Soc. Change*, 70 (2), 135–161.

O'Connor, G.C. Corbett, A.C. and Peters, L.S. (2018) *Beyond the Champion, Institutionalizing Innovation Through People*. Stanford University Press.

Ogilvy, J.A. (2022) *Creating Better Futures: Scenario Planning as a Tool for Better Tomorrow*. Oxford University Press.

Oneill, P. and Sohal, A.S. (1999) Business process re-engineering: A review of recent literature. *Technovation*, 19, 571–581.

Paluszny, A. and Nicholls, P.P. (1978) Predicting time dependent reliability of ceramic rotors. In *Ceramics for High Performance Applications*. Brook Hill.

Pararai, M., Warahena-Liyanage, G. and Oluyede, B.O. (2015) A new class of generalized power Lindley distribution with applications to lifetime data. *Theor. Math. Appl.*, 5 (1), 53–96.

Patherange, Y.L. et al. (2020) Organizational culture towards corporate performance. *J. Manag. Account. Econ.*, 7 (9), 522–549.

Patil, D., Rane, N. L. and Rane, J. (2024) Challenges In implementing Chat GPT and generative artificial intelligence in various business sectors. In *The Future Impact of Chat GPT on Several Business Sectors*. Deep Science Publishing, pp. 107–145.

Peter, T.J. and Waterman, R.H. (1982) *In Search of Excellence: Lessons from America's Best-run Companies*. Harper and Row.

Petrides, K.V., Furnham, A. and Mavroveli, S. (2007) Trait emotional intelligence: Moving forward in the field of EI. In G. Matthews, M. Zeidner and R.D. Roberts (eds.) *Emotional Intelligence: Knowns and Unknowns*. Series in Affective Sciences. Oxford University Press, pp. 151–166.

Pettigrew, A. (1985) *The Awakening Giant in ICI*. Oxford Blackwell.

Pfeffer, J. (1998) Seven principles of successful organizations. *California Manag. Rev.*, 40 (2), 96–124.

Pyzdek, T. (2001) *The Six Sigma Handbook*. McGraw Hill.

Rahman, M.M. and Watanabe, Y. (2023) Chat GPT for education and research: Opportunities, threats and strategies. *Appl. Sci.*, 13 (9), 5783.

Ranjibarian, M. (2022) A review of studies on organizational excellence. *Eurasian J. Sc. Tech.*, 2 (1), 44–52.

Rao, T.V. (2016) *Performance Management*. Sage.

Razeghi, A. (2008) *The Riddle*. Jossey Bass, pp. 1, 24.

Reigers, H.A. (2021) Business process management: The evolution of a discipline. *Comput. Ind.* https://doi.org/10.1016/j.compind.2021.103404

Ringrose, D. (2016) Organizational excellence. In A. Farazmand (ed.) *Global Encyclopaedia of Public Administration, Public Policy and Government*. Springer.

Russell, R.F. (2001) The role of values in servant leadership. *Leadership Org. Dev. J.*, 22 (2), 76–84.

Salovey, P. and Mayer, J.D. (1990) Emotional Intelligence: Imagination. *Cogn. Pers.*, 9, 185–221.

Sarhan, A.M. and Kundu, D. (2009). Generalized linear failure rate distributions. *Commun. Stat. Theor. Methods*, 38 (5), 642–660.

Savitz, A. (2006) *The Triple Bottom Line*. Jossey-Bass.

Schein, E.H. (1984, January 15) Coming to a new awareness of organizational culture. *Sloan Mgt. Rev.*, 25, 2.

Schneider, B. and Barbera, K.M. (2013) *The Oxford handbook of Organizational Climate and Culture*. Oxford University Press.

Schneider, B., Ehrhart, M.G. and Macey, W.H. (2013) Organizational climate and culture. *Ann. Rev. Psychol.*, 64, 361–388.

Schulze, R. and Roberts, R.D. (2005) *Emotional Intelligence: An International Handbook*. Hogrefe and Huber.

Shelton, C.D., Daling, J.K. and Walker, J.W. (2002) Foundations of organizational excellence: Leadership, values, strategies and skills. *LTA*, 1 (2), 48.

Shingo, S. (1986) *Zero Quality Control: Source Inspection and the Poka Yoke System.* Productivity Press.

Slaper, T.F. and Hall, T.J. (2011) The triple bottom line. What is it and how does it work? *Ind. Bus. Rev.*, 86 (1), 4–8.

Smith, R. (2024, November 8) The change diaries: Revealing the truth. Common myths in change management (blog). www.andchange.com

Sorgarde, W. (2023). Managing change in organizations. *Ann. Rev. Psych.*, 64, 361–388.

Spears, L.C. (2010) Character and servant leadership: Ten characteristics of effective, caring leaders. *J. Virtues Leadership*, 1, 25–30.

Stefan, A. (2021) Pursuing perfection or striving for excellence? 5 tips to achieve excellence. *Brainz Mag.* www.brainzmagazine.com

Sumanasiri, E.A.G. (2020) Value-based organizational leadership: A literature review. *J. Econ. Manag. Trade*, 26 (4), 92–104.

Sveningsson, S. and Sorgarde, W. (2023) *Managing Change in Organizations.* Sage.

Taylor, D.H. (2010) *The Power of Value-Based Leadership.* Author House Publishing.

Thabrew, L. et al. (2018) Using triple bottom line metrics and multi-criteria methodology in corporate settings. *J. Environ. Plan. Manag.*, 61 (1), 49–63.

Tierney, W. and Lanford, M. (2016) Conceptualizing innovation in higher education. In *Higher Education: Handbook of Theory and Research.* Springer, pp. 1–40.

Tijssen, R.J.W. and Winnink, J.J. (2017) Indicators, international statistics and innovative universities. *Scientometrics*, 114, 689–699.

Ubaid, A.M. and Dweiri, F.T. (2021) Business excellence models: A critical review of the models' frameworks. *Int. J. Bus. Excel.*, X(Y), 40–.

Ubaid, A.M. Dweiri, F.T. and Ojiako, U. (2020) Organizational excellence methodologies: A systematic literature review. *Int. J. Syst. Assur. Eng. Manag.*, 11 (1), 1395–1432.

Ulwick, A.W. (2005). *What Customers Want.* McGraw Hill.

UNESCO Ethics of Artificial Intelligence: The Recommendations. 2022. UNESCO.

Unnikrishnan, P.M., Trikone, J. and Agariya, A.K. (2019) Total quality management, two business excellence: A research journey. *Int. J. Bus. Excel.*, 19 (3), 323–363.

Viertl, R. (1988). *Statistical Methods in Accelerated Life Testing.* Vandenhoeck and Ruprecht.

Vilma, Z. (2018) Leadership values and values-based leadership: What is the focus? *App. Res. Health Soc. Sci.*, 15 (1), 43–58.

Vrcelj, N. (2023) Towards successful organizational change management: new process model identification. *J. Process Manag. New Technol.*, 11 (3–4), 1–15.

Webster, M. (2016) *Organizational Excellence, So What Is Organizational Excellence? A Blueprint for Success.* https://realorganizationalexcellence.com. Accessed August 25, 2011.

Westin, L. and Parmier, J. (2020) Inclusion of CSR in the extended performance satisfaction index—new developments. *Total Qual. Manag. Bus. Excel.*, 36 (3–4), 237–248.

Wilhelm, O. (2005) Measures of emotional intelligence; practice and standards. In R. Schulze and R.D. Roberts (eds.) *International handbook of Emotional Intelligence.* Hogrefe and Huber, pp. 131–154.

Zairi, M. (1997) Business process management: A boundary-less approach to modern competitiveness. *Bus. Process Manag. J.*, 3 (1), 64–80.

Zebec, A. and Spemberger, M.I. (2024) Creating artificial intelligence business value through BPM capabilities. *Bus. Process Manag. J.*, 30 (8), 1–26.

Index

T

U

V

W

X

Z

For Product Safety Concerns and Information please contact our EU representative GPSR@taylorandfrancis.com Taylor & Francis Verlag GmbH, Kaufingerstraße 24, 80331 München, Germany

Printed and bound by CPI Group (UK) Ltd, Croydon, CR0 4YY
14/09/2026
02216223-0016